The 2025 Suwon ITS Asia Pacific Forum

The 2025 Suwon ITS Asia Pacific Forum

Volume Editor

Sehyun Tak

MDPI

Basel • Beijing • Wuhan • Barcelona • Belgrade • Novi Sad • Cluj • Manchester

Volume Editor
Sehyun Tak
The Korea Transport Institute
Sejong-si
Republic of Korea

Editorial Office
MDPI AG
Grosspeteranlage 5
4052 Basel, Switzerland

This is a reprint of the Proceedings, published open access by the journal *Engineering Proceedings* (ISSN 2673-4591), freely accessible at: https://www.mdpi.com/2673-4591/102/1.

For citation purposes, cite each article independently as indicated on the article page online and as indicated below:

Lastname, A.A.; Lastname, B.B. Article Title. *Journal Name* **Year**, *Volume Number*, Page Range.

ISBN 978-3-7258-5235-2 (Hbk)
ISBN 978-3-7258-5236-9 (PDF)
https://doi.org/10.3390/books978-3-7258-5236-9

Cover image courtesy of Kim Hyeonrak

Contents

engineering proceedings

MDPI

Editorial

Statement of Peer Review

Sehyun Tak

The Korea Transport Institute, Sejong-si 30147, Republic of Korea; sehyun.tak@koti.re.kr

In submitting conference proceedings to *Engineering Proceedings*, the Volume Editors of the proceedings would like to certify to the publisher that all papers published in this volume have been subjected to peer review by the designated expert referees and were administered by the Volume Editors strictly following the policies.

The whole process was supervised by the conference committee of the forum and the Volume Editors and adhered to the professional and scientific standards expected of a proceedings journal published by MDPI and complied with the peer review policy and guidelines of *Engineering Proceedings*, which can be found at the following link: https://www.mdpi.com/journal/engproc/instruction_for_conference_organizers. The review reports were checked and archived by the Editorial Office of *Engineering Proceedings*.

- Type of peer review: single-blind
- Conference submission management system: 2025 Suwon ITS AP Forum Online Submission Platform in the official website.
- Number of submissions received: 188
- Number of submissions accepted: 15
- Acceptance rate (number of submissions accepted/number of submissions received): 8%
- Average number of reviews per paper: 2
- Total number of reviewers involved: 17

The Conference Committee extends our heartfelt thanks to the dedicated reviewers whose expertise and commitment have been key to the 2025 Suwon ITS AP Forum's success.

Conflicts of Interest: The author declares no conflict of interest.

engineering proceedings

Proceeding Paper

Understanding Commuters' Willingness to Shift to Transfer-Type Buses Using a Latent Class Model †

Hwan-Seung Lee and Ho-Chul Park *

Department of Transportation Engineering, Myongji University Yongin, Yongin 17058, Republic of Korea; hwanseung417@mju.ac.kr
* Correspondence: hcpark@mju.ac.kr; Tel.: +82-31-330-6507
† Presented at the 2025 Suwon ITS Asia Pacific Forum, Suwon, Republic of Korea, 28–30 May 2025.

Abstract

The Korean government proposes introducing a transfer-type bus system to reduce urban congestion. Transfer-type buses turn around at the Seoul border, requiring passengers to transfer to other modes to reach downtown. These buses have shorter routes, allowing reduced headways and increased bus supply. While this approach reduces congestion in the downtown area, it may cause transfer resistance, making it essential to analyze willingness to shift (WTS). This study uses a latent class model to categorize potential interregional bus users into three types: transfer avoidance, cost-sensitive, and time-sensitive. Over 50% of users in each group express WTS, showing a positive response to the transfer-type bus introduction. The choice model results indicate that the travel time and cost of direct type buses affect WTS, suggesting that policies should consider these factors for effective implementation.

Keywords: willingness to shift; latent class model; transfer-type bus; urban congestion

1. Introduction

As urban areas grow, the increase in commuting populations has led to significant congestion in downtown areas, especially in Seoul, South Korea, due to urban expansion. Efforts to alleviate congestion have included the implementation of congestion charges, but challenges persist with the high demand for interregional buses during peak times, leading to increased congestion and decreased competitiveness of public transportation [1]. In 2019, the Korean Ministry of Land, Infrastructure and Transport proposed a reorganization of the bus system, introducing direct-type buses, which enter downtown directly, and transfer-type buses, which require passengers to shift to other transportation modes at transfer centers located at the city borders [2]. This measure aims to reduce downtown congestion, shorten bus routes, and improve operation frequency and efficiency. However, the success of this initiative depends on passengers' willingness to shift (WTS) to this new system despite potential resistance due to the inconvenience of transferring. Therefore, it is important to analyze users' WTS before introducing the transfer-type bus system. Previous studies have discussed the concept and importance of WTS [3–5].

2. Methodology

This study was developed to estimate WTS by considering the latent preferences of interregional bus users. A stated preference (SP) survey was conducted with 502 interregional bus users in the Seoul Metropolitan Area (SMA), consisting of a latent preference survey

https://doi.org/10.3390/engproc2025102001

and a bus choice survey. Using the latent preference survey data, latent class analysis (LCA) was performed to classify users based on their latent behavioral preferences, and the WTS for each class was estimated. A choice model was then developed to compare direct-type buses, which enter downtown directly, and transfer-type buses, which require transfers at the city boundary. The model's reliability was verified.

Previous studies on WTS mainly relied on the binary logit model, which assumes that all individuals share the same utility function, limiting its ability to capture user diversity [6]. To address this, this study adopted the latent class model (LCM), which accounts for taste heterogeneity (TH) by assuming the existence of different latent classes with distinct choice utility functions [7]. The LCM consists of two components. The membership model uses LCA to categorize users into distinct classes based on shared characteristics and determine their probability of belonging to a specific class. The class-specific choice model estimates the probability of individuals selecting a transportation alternative within each latent class.

3. Results and Discussion

The three latent classes are estimated using LCA. Results show the probability of users belonging to each class based on their responses to the latent class survey questions. Class 1 was defined as the transfer-avoidance type because it exhibited negative characteristics related to transfer items and included 9.16% of users. Class 2 was defined as the cost-sensitive type due to its positive characteristics concerning cost items, comprising 41.24% of users. Class 3 was defined as the time-sensitive type as it showed common characteristics regarding time items, with 49.60% of users belonging to this class. The study analyzed the factors influencing the choice of transfer-type buses for each class of users, which are presented in Table 1.

Table 1. Estimation results of the model.

Coefficient			Parameter	Standard Error	*p*-Value
Class 1	Variables	X_{res}	−2.002	0.561	0.000
		X_{tp}	1.120	0.348	0.001
		$X_{ttb\_support}$	1.135	0.332	0.001
		X_{age_20}	0.908	0.233	0.000
		X_{ttc_Direct}	0.031	0.017	0.070
	Summary statistics	Cox and Snell R^2		0.133	
		Nagelkerke R^2		0.177	
		−2 Log-likelihood		456.632	
		Observation		368	
Class 2	Variables	X_{res}	0.307	0.183	0.094
		X_{income_low}	0.287	0.117	0.014
		X_{cbp_fare}	0.452	0.107	0.000
		X_{job_self}	0.836	0.327	0.011
		X_{ttc_Direct}	0.033	0.008	0.000
	Summary statistics	Cox and Snell R^2		0.029	
		Nagelkerke R^2		0.040	
		−2 Log-likelihood		2,109,358	
		Observation		1656	

Table 1. *Cont.*

Coefficient			Parameter	Standard Error	*p*-Value
Class 3	Variables	$X_{ttb\_support}$	0.588	0.130	0.000
		X_{cbp_fare}	0.186	0.098	0.058
		$X_{job\_student}$	0.720	0.172	0.000
		X_{ttc_Direct}	0.033	0.015	0.024
	Summary statistics	Cox and Snell R^2		0.028	
		Nagelkerke R^2		0.037	
		−2 Log-likelihood		2,686,118	
		Observation		1992	

For Class 1 (transfer-avoidance type), individuals were more likely to choose transfer-type buses if they lived in Seoul, made non-commute trips, supported the introduction of transfer-type buses, perceived high travel costs for direct-type buses, or were in their 20s. Despite their tendency to avoid transfers, they are willing to shift if they believe it will help reduce downtown congestion and if the cost of transfer-type buses is lower than that of direct-type buses. For Class 2 (cost-sensitive type), individuals were more likely to shift if they lived in the outskirts of Seoul, had a lower income (below USD 28,780), supported the new bus system, perceived high fares as a key issue, were self-employed, or found direct-type buses expensive. This class is highly sensitive to cost, which significantly influences their transportation choices. For Class 3 (time-sensitive type), higher willingness to shift was observed among students, those who use interregional buses to save time, individuals who support the transfer-type bus system, and those who perceive direct-type buses as too slow. This class prioritizes travel time, leading to the identification of time-related variables.

Author Contributions: Conceptualization, H.-C.P.; methodology, H.-S.L. and H.-C.P.; software, H.-S.L.; original draft preparation, H.-S.L.; writing—review and editing, H.-C.P. All authors have read and agreed to the published version of the manuscript.

Funding: This work was supported by 2024 Research Fund of Myongji University.

Institutional Review Board Statement: Not applicable.

Informed Consent Statement: Not applicable.

Data Availability Statement: The datasets used and/or analyzed during the current study available from the corresponding author on reasonable request.

Conflicts of Interest: The authors declare no conflict of interest.

References

1. Metropolitan Transport Commission. *A Study on the Diversification of Metropolitan Area Transportation Rate System;* Metropolitan Transport Commission: Sejong-si, Republic of Korea, 2020.
2. Korea Transport Institute. *A Study on the Development of Complex Transfer Centers by Transportation Axis in the Seoul Metropolitan Area;* Korea Transport Institute: Seoul, Republic of Korea, 2019.
3. Abouelela, M.; Al Haddad, C.; Antoniou, C. Are young users willing to shift from carsharing to scooter–sharing? *Transp. Res. Part D Transp. Environ.* **2021**, *95*, 102821. [CrossRef]
4. Halawani, A.T.; Rehimi, F. Evaluation of the intention to shift to public transit in Saudi Arabia. *Transp. Res. Part D Transp. Environ.* **2021**, *94*, 102809. [CrossRef]
5. Kwan, S.C.; Sutan, R.; Hashim, J.H. Trip characteristics as the determinants of intention to shift to rail transport among private motor vehicle users in Kuala Lumpur, Malaysia. *Sustain. Cities Soc.* **2018**, *36*, 319–326. [CrossRef]

6. Kim, S.H.; Choo, S. Developing a Latent Class Model Considering Heterogeneity in Mode Choice Behavior: A Case of Commuters in Seoul. *J. Korea Inst. Intell. Transp. Syst.* **2019**, *18*, 44–57. [CrossRef]
7. Bae, Y.G.; Jeong, J.H.; Kim, H.J. Latent Class Analysis for Mode Choice Behavior. *J. Korean Soc. Transp.* **2010**, *28*, 99–107.

engineering proceedings

Proceeding Paper

Development of High-Speed Rail Demand Forecasting Incorporating Multi-Station Access Probabilities †

Seo-Young Hong and Ho-Chul Park *

Department of Transportation Engineering, Myongji University, Yongin 17058, Republic of Korea; h583835@mju.ac.kr
* Correspondence: hcpark@mju.ac.kr
† Presented at the 2025 Suwon ITS Asia Pacific Forum, Suwon, Republic of Korea, 28–30 May 2025.

Abstract

This study develops a high-speed rail demand prediction model based on access probability, which quantifies the likelihood of passengers choosing a departure station among multiple alternatives. Traditional models assign demand to the nearest station or rely on manual calibration, often failing to reflect actual travel behavior and requiring excessive time and resources. To address these limitations, this study integrates survey data, real-world datasets, and machine learning techniques to model station choice behavior more accurately. Key influencing factors, including headway, access time, parking availability, and transit connections, were identified through passenger surveys and incorporated into the model. Machine learning algorithms improved prediction accuracy, with SHAP analysis providing interpretability. The proposed model achieved high accuracy, with an average error rate below 3% for major stations. Scenario analyses confirmed its applicability in network expansions, including GTX openings and the integration of mobility as a service. This model enhances data-driven decision-making for rail operators and offers insights for rail network planning and operations. Future research will focus on validating the model across diverse regions and refining it with updated datasets and external data sources.

Keywords: transportation demand forecasting; high-speed rail; multi-station access probability; travel behavior; machine learning; data-driven; multi-layer perceptron (MLP)

1. Introduction

Traffic demand forecasting plays a critical role in railway investment and operational planning, yet traditional four-step models fail to accurately reflect passenger behavior. These models typically assign demand to the nearest station within an administrative district, disregarding alternative station choices. Studies show that only 47% of railway users select the closest station; among multimodal travelers, this figure drops to 40% [1,2]. To address this, network calibration is often applied, but it is resource-intensive and reliant on subjective adjustments. Research highlights the importance of access time in station choice, with one study finding it nearly twice as influential as in-vehicle time [3]. However, existing studies primarily focus on identifying influencing factors rather than integrating station choice and demand estimation into a unified forecasting model. Recent advancements in big data, automation, and machine learning offer promising solutions for improving demand forecasting accuracy. Machine learning models have demonstrated superior predictive performance compared to traditional logit-based approaches [4]. However, their lack of interpretability remains a challenge, necessitating methods like SHAP (Shapley

 https://doi.org/10.3390/engproc2025102002

Additive Explanation) to enhance model transparency [5]. This study aims to develop a machine learning-based demand forecasting model by estimating access probabilities, overcoming the limitations of distance-based assignment models. By incorporating access probabilities into station choice modeling, the proposed framework enhances prediction accuracy and provides a more precise station-level demand estimation method for high-speed rail networks. Its performance will be validated through real-world transportation scenarios to ensure practical applicability.

2. Data & Methodology

This study models access probability to enhance high-speed rail (HSR) demand forecasting, estimating the likelihood of passengers selecting a departure station (Figure 1). Given that the exact demand distribution across stations is unknown, an optimization technique is applied, leveraging station-specific demand data and inter-zone rail traffic volumes. The analysis identifies three alternative stations per zone based on centroid distances, ensuring a realistic representation of station choice behavior. The estimated access probabilities serve as outputs in the MLP model, while station-specific characteristics act as input variables to refine demand predictions.

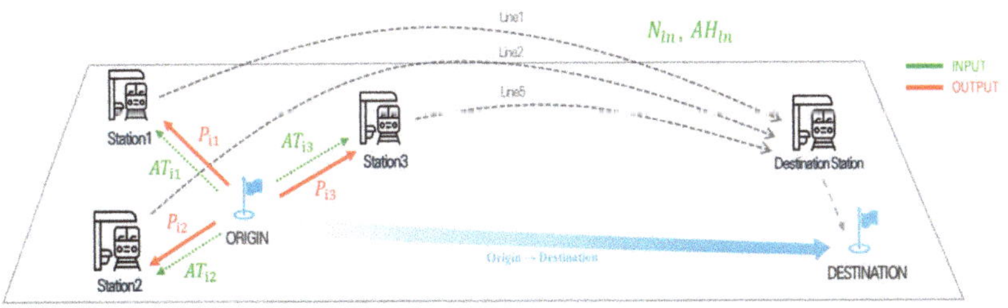

Figure 1. The concept of access probability in OD trips.

To support this estimation, the study utilizes 2019 KTDB O-D data, focusing on five major rail lines, connecting 12 key departure stations in the Seoul metropolitan area to 1135 zones nationwide. Each zone considers three alternative stations, resulting in 16,566 observations. The dataset integrates station accessibility, service frequency, and operational factors, including access time (TMAP API, smart card data), train headways (timetables), parking capacity, and transit connections. These variables, sourced from 2021–2023 datasets, account for regional and station-specific differences, forming a robust foundation for access probability estimation and improving demand forecasting accuracy.

This study employs a multi-layer perceptron (MLP) model, a type of supervised artificial neural network (ANN), to enhance high-speed rail (HSR) demand forecasting by capturing complex, non-linear relationships between input variables and station choice behavior. The model consists of three layers: an input layer, two hidden layers (with 8 and 6 units, respectively), and an output layer. Each node adjusts its activation through backpropagation, minimizing errors using the Levenberg–Marquardt training algorithm. The ReLU function is used for hidden layer activation, while Softmax represents station choice probabilities in the output layer.

The dataset was split into 70% training and 30% testing sets, with feature scaling applied using Standard Scaler to normalize input values. The model was implemented using Keras, trained with Mean Squared Error (MSE) as the loss function and optimized using Adam optimizer. The batch size was set to 32, with a learning rate of 0.001, and the model was trained for 10 epochs. Performance was assessed through the Area Under the Curve (AUC) metric, ensuring robust predictive accuracy.

3. Results and Discussion

The optimization process effectively distributed trip volumes across three alternative stations per zone, with access probabilities validated against Railway Statistical Yearbook data, showing an error rate close to 0%. The MLP model demonstrated a strong predictive performance (Table 1), achieving a Mean Squared Error (MSE) of 0.0214, a Mean Absolute Error (MAE) of 0.0990, and a coefficient of determination (R^2) of 0.7529. Errors were below 1% for major stations (Seoul, Yeongdeungpo, Suseo, Gwangmyeong, Yongsan) and within 3% for Dongtan, Haengsin, and Sangbong, though stations with low demand (<5000 trips) showed higher errors, indicating limitations in sparsely used locations. SHAP analysis identified train frequency, bus routes, headway, and parking capacity as key factors influencing station choice, with lower train frequency and longer headways reducing access probability.

Table 1. MLP estimation results.

Station	Actual Value	Forecasted Value		
		Estimated Value	Difference	Error Rate
Seoul	85,022	85,011	−11	0%
Suseo	41,438	41,086	−352	−1%
Gwangmyeong	27,581	27,676	95	0%
Yongsan	25,764	25,782	18	0%
Dongtan	8249	8475	226	3%
Cheongnyangni	5055	5980	925	18%
Haengsin	4291	4170	−121	−3%
Pyeongtaek/Jije	3741	4592	851	23%
Suwon	3686	2861	−825	−22%
Yeongdeungpo	1320	1315	−5	0%
Sangbong	977	1010	33	3%
Yangpyeong	839	5	−834	−99%

The case study analysis assessed the impact of key variable changes on access probability and rail demand, demonstrating the model's applicability in real-world scenarios. One scenario analyzed the effect of mobility as a service (MaaS) on transit efficiency, showing that reducing public transit access time by 10–30% led to a 2–5% increase in access probability, with greater improvements observed in peripheral areas.

This study introduces a novel approach to rail demand forecasting, addressing the limitations of traditional models while enhancing predictive accuracy. By incorporating access probability, the proposed method provides valuable insights for optimizing station accessibility and supporting data-driven policy and network planning.

Author Contributions: The authors confirm the following contributions to this paper: study conception and design: S.-Y.H. and H.-C.P.; data collection: S.-Y.H.; analysis and interpretation of results: S.-Y.H. and H.-C.P.; and draft manuscript preparation: S.-Y.H. and H.-C.P. All authors have read and agreed to the published version of the manuscript.

Funding: This research was supported by the National Research Foundation of Korea (NRF) grant funded by the Korean government (MSIT) (RS-2024-00348596).

Institutional Review Board Statement: Not applicable.

Informed Consent Statement: Not applicable.

Data Availability Statement: The data presented in this study are available upon request from the corresponding author.

Conflicts of Interest: The authors declare no conflict of interest.

References

1. Debrezion, G.; Pels, E.; Rietveld, P. Modelling the Joint Access Mode and Railway Station Choice. *Transp. Res. Part E Logist. Transp. Rev.* **2009**, *9*, 270–283. [CrossRef]
2. Cheon, M.J.; Choi, H.J.; Park, J.W.; Choi, H.Y.; Lee, D.H.; Lee, O. A Study on the Traffic Prediction through CatBoost Algorithm. *J. Korea Acad.-Ind. Coop. Soc.* **2021**, *22*, 58–64.
3. Lee, J. A Development of Intercity Travel Mode Choice Model for High-Speed Rail Demand Analysis. *J. Transp. Res. Korean Soc. Transp.* **2009**, *16*, 27–40.
4. Zhang, X.; Zhao, X. Machine Learning Approach for Spatial Modeling of Ride sourcing Demand. *J. Transp. Geogr.* **2022**, *100*, 103310. [CrossRef]
5. Lundberg, S.M.; Lee, S.I. A Unified Approach to Interpreting Model Predictions. In Proceedings of the 31st Conference on Neural Information Processing Systems (NeurIPS 2017), Long Beach, CA, USA, 4–9 December 2017; pp. 4766–4777.

engineering proceedings

MDPI

Proceeding Paper

Influence of Selective Security Check on Heterogeneous Passengers at Metro Stations †

Zhou Mo, Maricar Zafir and Gueta Lounell Bahoy *

Hitachi Asia Ltd., 20 Pasir Panjang Road #05-21, Singapore 117439, Singapore
* Correspondence: lb.gueta.vx@hitachi.com; Tel.: +65-9180-3816

† Presented at the 2025 Suwon ITS Asia Pacific Forum, Suwon, Republic of Korea, 28–30 May 2025.

Abstract

Security checks (SCs) at metro stations are regarded as an effective measure to address the heightened security risks associated with high ridership. Introducing SCs without exacerbating congestion requires a thorough understanding of their impact on passenger flow. Most existing studies were conducted where SCs are mandatory and fixed at certain locations. This study presents a method for advising the scale and placement for SCs under a more relaxed security setting. Using agent-based simulation with heterogeneous profiles for both inbound and outbound passenger flow, existing bottlenecks are first identified. By varying different percentages of passengers for SCs and locations to deploy SCs, we observe the influence on existing bottlenecks and suggest a suitable configuration. In our experiments, key bottlenecks are identified before tap-in fare gantries. When deploying SCs near tap-in fare gantries as seen in current practices, a screening percentage of beyond 10% could exacerbate existing bottlenecks and also create new bottlenecks at SC waiting areas. Relocating the SC to a point beyond the fare gantries helps alleviate congestion. This method provides a reference for station managers and transport authorities for balancing security and congestion.

Keywords: metro; security check; congestion; heterogeneous passenger; agent-based simulation

1. Introduction

Urban rail transit systems, as the backbone of public transportation systems, are essential for daily commuting and connectivity in major cities. As commuting, tourism and entertainment activities return to pre-pandemic level, major metro stations are experiencing congestion—particularly during peak hours and around major public events such as service disruption, concerts and celebrations. Such crowding not only disrupts commuters' travel efficiency, but also heightens security risks. To address these concerns, security checks (SCs) are considered necessary at metro stations, despite impacting passenger efficiency.

Globally, there are mainly three types of SCs at metro stations depending on passenger coverage:

1. Comprehensive: All passengers need to go through door frame metal detectors and their bags need to be checked with handheld metal detectors or X-ray machines.
2. Selective: Passengers are randomly checked at selected stations during regular emergency preparedness exercises, major events, or sometimes daily operation.
3. Minimal: In this type, passengers are rarely checked upon entering metro stations, hence minimizing the impact on passenger flow.

Due to the varied adoption of SC practices worldwide, there have been limited studies investigating their impact on passenger flow. Most research has been conducted under the comprehensive SC mode. Some researchers focused on enhancing SC efficiency by optimizing facility configurations. Yu et al. modelled passenger flow in a station in Guangzhou using a simulation and found that existing SC facilities cannot meet the requirements of passenger flow in extreme peak hours [1]. They proposed measures such as adding SC machines and ticket vending machines and saw an improvement in crowd density and the number of inbound passengers. Wei et al. suggested an optimal X-ray conveyor belt length to reduce waiting time at SC queues based on the M/M/1/N queuing model [2]. In a subsequent study [3], they recommended suitable combinations of X-ray machines and detector doors that yield the shortest systematic and individual waiting time and least serious congestion. Other studies explored expediting the SC process, such as by adopting a dedicated SC channel [3–5] for passengers with small bags or skipping SCs for passengers without bags [6]. These countermeasures have been proven effective in reducing crowd density in areas prior to fare gantries, increasing served passengers and shortening SC waiting time. However, there are still some gaps left to be addressed:

1. Most existing studies are conducted in mainland China under the comprehensive SC type. This might be difficult to directly reference for other places where an SC is not a norm, especially on how many passengers should be screened to avoid congestion.
2. Since SC facilities in these studies tend to be bulky, the locations for SCs are usually fixed (at station entry points). There is one recent study [7] that considered moving SCs from the station hall to the entrance channel, which is sufficiently long for congestion prevention. However, this might not apply for stations without long entrance channels.
3. Most studies only simulate inbound passengers or passenger flow near SCs. Only [7] has modelled alighting, boarding and transfer passengers when identifying bottlenecks.
4. While some studies differentiate passenger profiles by luggage, the impact of such heterogeneity on their behaviours is considered only in the context of SC channel selection. More fine-grained behaviours such as passenger sizes, walking speeds and preferences for facilities in the stations were not modelled accordingly.

This study proposes a method to evaluate the impact of SCs on heterogeneous passengers at metro stations where an SC is non-mandatory. An agent-based simulation model with fine-grained passenger profiles and behaviours was built to identify bottleneck areas. By exploring screening percentages of different profiles and varying locations of SCs, we evaluated the influence on key bottleneck areas and suggested a suitable scale and placement of SCs with the least disruption. This method serves as a guidance to station managers or transport authorities for SC deployment.

2. Methodology

2.1. Simulation Environment

In this study, we build a microscopic passenger flow model for a metro station by using AnyLogic software (version 8.9.3), which is widely adopted in research reviewed in the previous section. To demonstrate the proposed method, the station layout, facilities, and passenger statistics in City Hall station in Singapore were taken as an example. It is an underground cross-platform interchange station in the central region of Singapore, connecting two Mass Rapid Transit (MRT) lines, both operating high-capacity 6-car trainsets. The station provides access to shopping malls, office buildings, tourist attractions and public events such as the F1 race, National Day Parade and New Year's Eve celebrations. During the events, random screenings were conducted.

Figure 1 shows the station layout and facilities. The station has three levels, the top concourse level (denoted as B1) and platform levels below (denoted as B2 and B3), arranged in two stacked island platforms. The green escalators on both sides connect B1 and B2 and the yellow escalators, near the elevator, link B1 and B3.

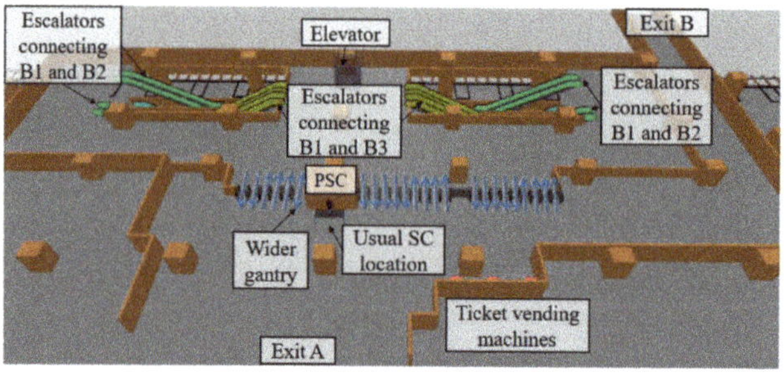

Figure 1. Concourse layout and facilities.

Based on the farecard transaction data collected over 22 working days, as shared by [8], the average passenger flow at the station peaks between 6 and 7 pm, with 6391 passengers tapping in and 5487 passengers tapping out. In the simulation, we assign these inflow passengers equally to both exits and outflow passengers to trains with a peak headway of 3 min.

There are three sets of fare collection gantries aligned with the passenger service centre (PSC). We adopted an afternoon peak gantry direction setting as shown by the blue arrows in Figure 1, where 12 gantries are dedicated to tapping out and 11 gantries for tapping in. A bidirectional, wider gantry is placed on the left of the PSC for passengers with personal mobility devices (PMDs) such as wheelchairs.

2.2. Passenger Profiles and Behaviour Models

To represent real-life passengers, we defined seven passenger profiles categorized by age group, PMD usage and luggage usage: middle-aged adults (hereafter referred as "adults") without luggage, middle-aged adults with luggage, elderly adults (hereafter referred as "elders") without luggage, elderly adults with luggage, children without luggage, children with luggage, and PMD users. We assume the percentage of elders and children and PMD users to be 10%, 10% and 1%, respectively, while the rest are adults. Within each profile, 5% are laden with luggage. Table 1 shows the passenger profiles and their parameters. The embedded social force model [6] is used to model passenger flow. The speed and body diameter of non-PMD users were based on references [9,10]. The speed of PMD users was based on [11] and the diameter was set in accordance with dimension restriction by local authorities [12]. For luggage-laden passengers, the walking speed is reduced by 30% [13] and the diameter is increased by 0.17 m [14] compared to those without luggage.

We assume all passengers except PMD users prefer the gantry with the shortest queue. If gantries have the same queue lengths, passengers chose the nearest gantry. Further, PMD users choose the wider gantry. Since it is bi-directional, the queue length in this gantry is counted by summing both sides. The time taken for tapping at fare gantries was assumed to follow a uniform distribution between 1 s and 2 s based on our field observation.

PMD users can only choose elevators due to their mobility restrictions. For other passengers, the probability of choosing elevators varies with elders having a higher probability, followed by adults and children. Such a probability is increased if the passenger carries luggage. The probability for elders, adults and children without luggage is assumed to be

10%, 5%, and 2.5%, which is doubled when the passenger is laden with luggage. They will choose escalators instead if their waiting time for elevators exceeds 30 s. The elevator has a capacity of 15 people, operating at 1 m/s. The escalators travel at 0.75 m/s during peak hours as per local guidelines.

Table 1. Passenger profiles with parameters.

	Profile	Comfortable Speed (m/s)	Diameter (m)
1	Adults without luggage	v_1~N (1.27, 0.14)	d_1~U (0.44, 0.58)
2	Adults with luggage	$v_1 \times$ (1–30%)	d_1 + 0.17
3	Elders without luggage	v_3~N (1.04, 0.21)	d_3~U (0.46,0.54)
4	Elders with luggage	$v_3 \times$ (1–30%)	d_3 + 0.17
5	Children without luggage	v_5~N (1.08, 0.23)	d_5~U (0.39, 0.45)
6	Children with luggage	$v_5 \times$ (1–30%)	d_5 + 0.17
7	PMD users	v_7~N (0.48, 0.21)	1.2

Based on field observations, most passengers inspected by security officers carry luggage. To facilitate interpretation, we assume a screening probability of 1 for all luggage-laden passengers and 0 for passengers without luggage. In other words, the screening percentage is equivalent to the percentage of passengers with luggage. The time taken for SCs was assumed to follow a uniform distribution from 7 s to 8 s [4]. More refined risk modelling is underway as future study.

Although passengers can pay travel fares using bank cards, some still use stored-value cards, which necessitates a visit to ticket vending machines for top-ups before proceeding to fare gantries. Given the high traffic volume during peak hours, which may discourage passengers from topping up, we assume these passengers constitute 1% of the simulated population.

2.3. Experimental Design

We initially excluded SCs in the experiment to identify bottleneck areas at the station concourse by observing the crowd density heatmap. Subsequently, SCs are added at the usual location to evaluate their impact in these bottleneck areas. Finally, we experiment on placing SCs at an unconventional location and evaluate congestions caused by SCs by comparing their effect on the bottleneck areas. All experiments are run in Intel (R) Core (TM) i7-1260P 2.10 Ghz Lenovo PC with 16 GB RAM sourced from Singapore.

3. Results and Discussion

3.1. Existing Bottlenecks with and Without SC

Figure 2 shows a heatmap visualizing the average crowd density without SCs. It is at a logarithmic scale to reflect small variations. As highlighted by areas in yellow and orange, key bottleneck areas occur before tap-in gantries, especially at the right gantry set due to the narrower path. This is probably due to the higher rate of inflow than outflow at fewer gantries. The dense areas near the escalators are caused by merging passengers from both directions, which can be solved by setting guardrails for separation [7]. The noticeably longer queue in front of the wider gantry is likely caused by the slower speed of PMD users and the delay when faced with passengers waiting in the tap-out direction.

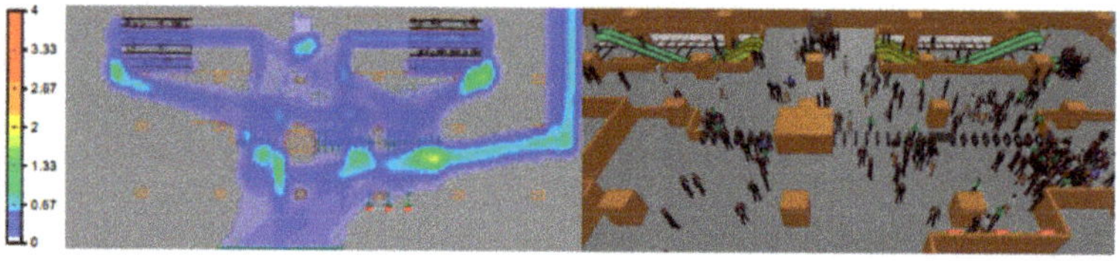

Figure 2. Crowd density without SC.

Figure 3 shows the heatmap after setting up a SC at its usual location. The red rectangle highlights the crowd in the SC queuing areas, which is still manageable under 1.33 pax/m^2. The crowd density at the right gantry set is increased and widened towards the middle gantry set due to the increased passengers with paths to the SC location.

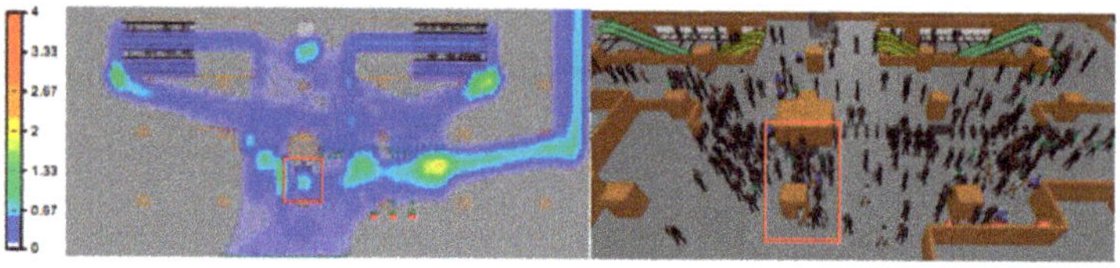

Figure 3. Crowd density with 5% SC.

3.2. *Influence of SC Percentage*

Figure 4 shows the congestion heatmap after increasing the SC percentage to 10% and 15%. Significant spillovers to the adjacent gantry sets are observed from 10%, making the queue length of the wide tap-in gantry longer. This causes a greater issue when increasing the percentage to 15%. The SC waiting area reached a higher density beyond 2.67 pax/m^2 at a larger area, creating a new bottleneck. Because of the blockage, some passengers exiting from the wide tap-out gantry experienced difficulty in reaching Exit A (shown in the red dotted line).

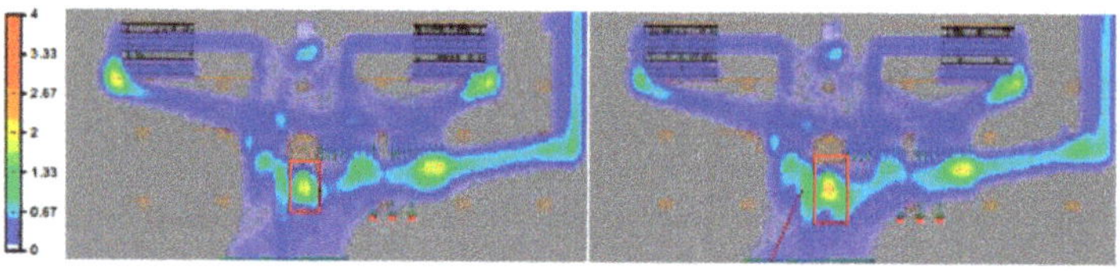

Figure 4. Crowd density with 10% (**left**) and 15% (**right**) SC.

The results suggest that for the current simulation layout and experiment setting, the optimal SC percentage that avoids causing additional congestion is between 5% and 10%. If the percentage exceeds 10%, other SC locations may need to be considered.

3.3. *Influence of SC Location*

Given the impact of SCs on the non-paying areas, we explored moving SCs to after the fare gantries. Inbound passengers are screened right after they tapped in, without increasing travel distance drastically. Figure 5 shows the comparison under 5% and 10% of the SC percentage. In both cases, the average crowd density in the SC waiting area reduces. This is due to the smaller volume of tap-out passengers at the gantries near the PSC.

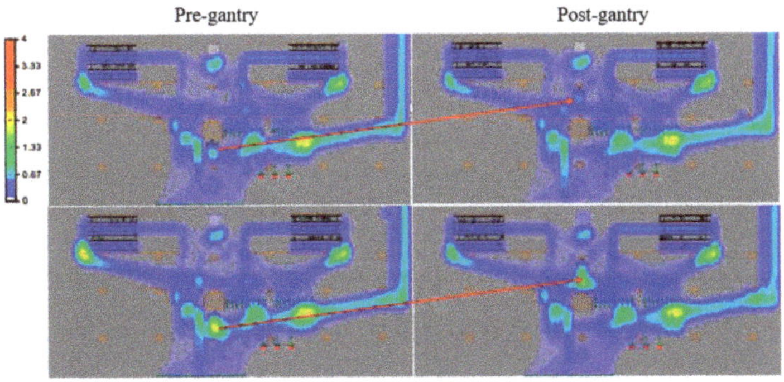

Figure 5. Crowd density with 5% (**top**) and 10% (**bottom**) SC pre- and post-gantry.

If the SC percentage is increased for security concerns, other locations may be considered for SCs. These areas can be at the end of corridor connecting Exit B and corner areas near the escalators connecting to B2. Both may help slow down passengers arriving at existing bottlenecks such as the one in front of the right tap-in gantry set and the escalator clusters, thus reducing crowd density and increasing passenger travel time at the concourse. Identifying the optimal location to balance the trade-off between reducing crowd density and minimizing passenger travel time is one of the key future directions of this ongoing research.

4. Conclusions

This study presents an agent-based modelling methodology to evaluate the influence of the SC scale and location on metro station congestion under heterogeneous passenger profiles. By comparing its influence on existing bottlenecks, we advise a 5–10% SC percentage at the current SC location and suggest exploring post-gantry SCs if the station managers wish to further alleviate congestion at the SC queuing area.

Future research directions include verifying the method at more stations with real-life data, exploring other underutilized areas for SCs, incorporating more diverse congestion evaluation metrics, and changing the SC strategy from by luggage to by risk. Risk modelling can leverage expert experience and be integrated with a computer-vision-based risk detection system at station entrances for enhanced accuracy, facilitating more efficient security screening.

Author Contributions: Conceptualization, Z.M.; methodology, Z.M. and G.L.B.; software, M.Z. and Z.M.; validation, Z.M. and G.L.B.; formal analysis, Z.M.; investigation, Z.M.; resources, Z.M.; data curation, Z.M.; writing—original draft preparation, Z.M.; writing—review and editing, G.L.B. and Z.M.; visualization, Z.M. and M.Z.; supervision, G.L.B.; project administration, G.L.B.; funding acquisition, G.L.B. All authors have read and agreed to the published version of the manuscript.

Funding: This research received no external funding.

Institutional Review Board Statement: Not applicable.

Informed Consent Statement: Not applicable.

Data Availability Statement: Not applicable. The data are not available for sharing.

Conflicts of Interest: The authors are employees of the company Hitachi Asia Ltd. and conducted this research as part of their roles in the company's Research & Development Centre. The authors declare no other conflicts of interest.

References

1. Yu, H.; Wang, Y.; Wang, F.; Qiu, P. Understanding impacts of security check on passenger flow in a metro station and improving measures: A case study in Guangzhou, China. *J. Adv. Transp.* **2019**, *1*, 7438545. [CrossRef]
2. Wei, Z.; Chu, S.; Huang, Z.; Qiu, S.; Zhao, Q. Optimization design of X-ray conveyer belt length for subway security check systems in Beijing, China. *Sustainability* **2020**, *12*, 2133. [CrossRef]
3. Wei, Z.; Liang, J.; Qiu, S.; Wang, S.; Liu, S. How Many Facilities Are Needed? Evaluating Configurations of Subway Security Check Systems via a Hybrid Queueing Model. *IEEE Trans. Intell. Transp. Syst.* **2021**, *23*, 8209–8222. [CrossRef]
4. Zhou, Y.; Zhao, M.; Sun, L. Optimization of bottleneck facilities in subway stations based on WiFi data. In *CICTP 2019: Transportation in China—Connecting the World*; Zhang, L., Ma, J., Liu, P., Zhang, G., Eds.; American Society of Civil Engineers: Reston, VA, USA, 2019; pp. 6287–6298.
5. He, B.; Liu, Y.; Gao, X.; An, F.; Lv, X. Passenger Queuing Analysis Method of Security Inspection and Ticket-Checking Area without Archway Metal Detector in Metro Stations. *Promet–Traffic Transp.* **2023**, *35*, 772–785. [CrossRef]
6. Wan, M.; Chen, Z.; Guo, J.; Wan, P. Optimization of security check efficiency in subway station based on Anylogic: A case study of Nanchang Metro. *J. Intell. Fuzzy Syst.* **2024**, *41*, 5035–5043. [CrossRef]
7. Peng, J.; Wei, Z.; Li, J.; Guo, X.; Wang, S. Passenger flow bottleneck decongestion in subway stations: A simulation study. *Simulation* **2024**, *100*, 981–995. [CrossRef]
8. E, J.; Li, M.; Huang, J. CrowdAtlas: Estimating crowd distribution within the urban rail transit system. In Proceedings of the 2021 IEEE 37th International Conference on Data Engineering (ICDE), Chania, Greece, 19–22 April 2021; pp. 2219–2224.
9. Yeo, S.K.; He, Y. Commuter characteristics in mass rapid transit stations in Singapore. *Fire Saf. J.* **2009**, *44*, 183–191. [CrossRef]
10. Lei, W.; Li, A.; Gao, R.; Hao, X.; Deng, B. Simulation of pedestrian crowds' evacuation in a huge transit terminal subway station. *Phys. A Stat. Mech. Its Appl.* **2012**, *391*, 5355–5365. [CrossRef]
11. Sonenblum, S.E.; Sprigle, S.; Lopez, R.A. Manual wheelchair use: Bouts of mobility in everyday life. *Rehabil. Res. Pract.* **2012**, *2012*, 753165. [CrossRef] [PubMed]
12. An Inclusive Public Transport System. Available online: https://www.lta.gov.sg/content/ltagov/en/getting_around/public_transport/a_better_public_transport_experience/an_inclusive_public_transport_system.html (accessed on 15 January 2025).
13. Ali, M.F.M.; Abustan, M.S.; Talib, S.H.A.; Abustan, I.; Abd Rahman, N.; Gotoh, H. A case study on the walking speed of pedestrian at the bus terminal area. In Proceedings of the E3S Web of Conferences, Penang, Malaysia, 28–29 November 2017; pp. 1–7.
14. Davis, D.G.; Braaksma, J.P. Adjusting for luggage-laden pedestrians in airport terminals. *Transp. Res. Part A Gen.* **1988**, *22*, 375–388. [CrossRef]

engineering proceedings

MDPI

Proceeding Paper

Beyond the Red and Green: Exploring the Capabilities of Smart Traffic Lights in Malaysia †

Mohd Fairuz Muhamad@Mamat *, Mohamad Nizam Mustafa, Lee Choon Siang, Amir Izzuddin Hasani Habib * and Azimah Mohd Hamdan

Highway Planning Division, Ministry of Works, Kuala Lumpur 50580, Malaysia
* Correspondence: sub_bpj@kkr.gov.my (M.F.M.); amirih@jkr.gov.my (A.I.H.H.)
† Presented at the 2025 Suwon ITS Asia Pacific Forum, Suwon, Republic of Korea, 28–30 May 2025.

Abstract

Traffic congestion poses a significant challenge to modern urban environments, impacting both driver satisfaction and road safety. This paper investigates the effectiveness of a smart traffic light system (STL), a solution developed under the Intelligent Transportation System (ITS) initiative by the Ministry of Works Malaysia, to address these issues in Malaysia. The system integrates a network of sensors, AI-enabled cameras, and Automatic Number Plate Recognition (ANPR) technology to gather real-time data on traffic volume and vehicle classification at congested intersections. This data is utilized to dynamically adjust traffic light timings, prioritizing traffic flow on heavily congested roads while maintaining safety standards. To evaluate the system's performance, a comprehensive study was conducted at a selected intersection. Traffic patterns were automatically analyzed using camera systems, and the performance of the STL was compared to that of traditional traffic signal systems. The average travel time from the start to the end intersection was measured and compared. Preliminary findings indicate that the STL significantly reduces travel times and improves overall traffic flow at the intersection, with average travel time reductions ranging from 7.1% to 28.6%, depending on site-specific factors. While further research is necessary to quantify the full extent of the system's impact, these initial results demonstrate the promising potential of STL technology to enhance urban mobility and more efficient and safer roadways by moving beyond traditional traffic signal functionalities.

Keywords: smart traffic light (STL); intelligent transportation system (ITS); traffic flow optimization; dynamic signal timing

1. Introduction

Traffic congestion is a persistent urban challenge, causing extended travel times, increased fuel consumption and air pollution, and higher accident risks. These issues have significant economic, environmental, and social consequences, demanding innovative solutions. Globally, congestion's financial burden reaches billions of dollars annually, stemming from lost productivity and wasted fuel. Malaysia alone estimated RM13.09 billion in congestion-related costs in 2015, including lost income and environmental damage. Rapidly urbanizing cities like Kuala Lumpur face particularly acute congestion due to increased vehicle ownership. Intelligent Transportation Systems (ITS) offer a promising approach by using technology to optimize traffic flow, improve safety, and enhance air quality. This paper, however, focuses on a key ITS component: Smart Traffic Lights (STLs). We explore STL capabilities that extend beyond traditional "red and green" signal systems.

This research investigates an STL implemented under Malaysia's National Intelligent Transportation Management Center (NITMC) initiative, a Ministry of Works project aimed at alleviating congestion and improving road safety. Unlike traditional traffic lights, which rely on preset timings and often fail to adapt to dynamic traffic conditions, the STL utilizes a sophisticated approach. Integrating sensors, AI-powered cameras, and Automatic Number Plate Recognition (ANPR) technology, the STL gathers and analyzes real-time traffic data. This enables dynamic optimization of signal timings, prioritizing flow based on immediate needs. This paper contributes to the understanding of ITS in traffic management by analyzing the STL's implementation and its impact on travel times and overall flow. This case study highlights the benefits and potential of STLs, demonstrating their viability for wider deployment. Ultimately, this research shows how STLs are transforming urban mobility, moving beyond simple red and green light sequencing towards a more dynamic, intelligent, and responsive system.

2. Literature Review

Traffic signal control has evolved significantly since early systems like the Sydney Coordinated Adaptive Traffic (SCAT) system [1]. SCAT's pioneering vision anticipated the potential for systems to learn from traffic, adapt dynamically, and identify vehicles—concepts that became foundational to modern Intelligent Transportation Systems (ITS). This emphasis on real-time adaptation remains central to ITS research. Subsequent work has explored Adaptive Traffic Control Systems (ATCS), connected vehicle technologies, and real-time data analysis to optimize traffic signals, reduce delays, and enhance network performance [2,3]. These advancements build upon the foundation laid by early innovators. However, complex traffic scenarios, especially those involving emergency vehicles, present ongoing challenges. Prioritizing emergency vehicle movement while minimizing disruption requires solutions beyond traditional signal timing. Qin and Khan [4], offer valuable insights, proposing control strategies using real-time data and optimization algorithms to improve emergency vehicle signal pre-emption (EVSP) effectiveness. This exemplifies smart traffic management's ability to address specific, real-world problems, moving beyond the limitations of older systems. In Malaysia, government policies actively support ITS development and implementation. The "Malaysian ITS Blueprint 2019–2023" and the "National Transport Policy 2019–2030" highlight ITS's strategic importance in modernizing the nation's transportation sector [5,6]. Kuala Lumpur City Hall (DBKL) explores technological solutions, including ITS, to address urban congestion [7]. The Malaysian Institute of Road Safety Research (MIROS) has developed guidelines for Tidal Flow Systems, demonstrating a commitment to innovative traffic management [8]. Furthermore, the establishment of the Malaysia National AI Office (NAIO) and the National Artificial Intelligence Roadmap 2021–2025 (AI-RMAP) signals a strong commitment to fostering a robust AI ecosystem. This focus on AI directly aligns with the technological underpinnings of ITS, particularly the AI-driven capabilities of STL. The NAIO will play a key role in promoting AI-powered solutions like STLs within Malaysia's infrastructure [9,10]. This study contributes to this progress by evaluating STL effectiveness in addressing traffic congestion, specifically its capabilities beyond traditional systems. By examining the STL's real-time adaptability and impact on traffic flow, this research demonstrates how these advanced systems are moving beyond basic traffic regulation towards a future of dynamic, intelligent, and responsive urban mobility in Malaysia.

3. Methodology

This study employed a mixed-methods approach, combining quantitative data analysis with visual observations to comprehensively understand the STL's impact and its capabilities beyond the limitations of conventional traffic signal control. The STL was deployed at strategically selected intersections in State of Terengganu, Kedah and Selangor to demonstrate its effectiveness in real-world scenarios.

3.1. System Architecture and Data Collection

The STL deployed in all locations moves beyond simple red and green signalization by employing a sophisticated architecture designed for real-time traffic management. This architecture comprises the following:

1. **A Network of Sensors:** Strategically positioned sensors were used to capture granular data on traffic volume and vehicle speed.
2. **AI-Powered Cameras:** These cameras were deployed to monitor and analyze traffic patterns, providing a comprehensive understanding of traffic dynamics beyond basic vehicle counts. This included identifying vehicle types and recognizing unusual traffic patterns.
3. **Automatic Number Plate Recognition (ANPR) Technology**: Integrated ANPR technology enables vehicle identification and tracking, contributing to a more detailed understanding of traffic flow and individual vehicle movements.

This seamless integration of sensors, AI, and ANPR facilitated the collection and processing of extensive real-time data, forming the foundation for the system's adaptive capabilities. Figure 1 below is the smart traffic light's system architecture.

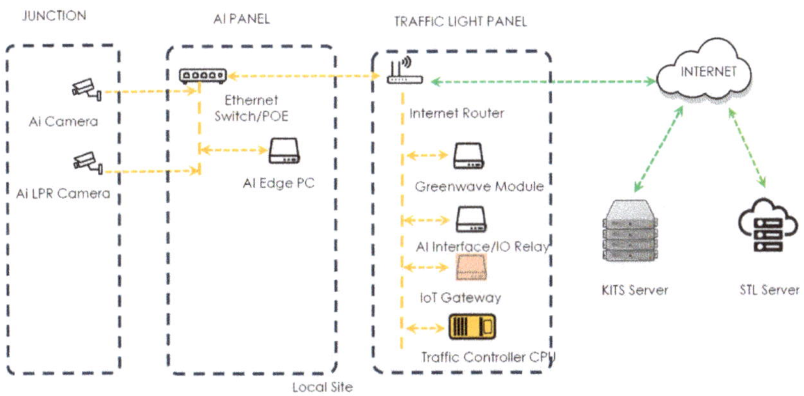

Figure 1. Smart traffic light's system architecture.

3.2. Study Location

The STL was implemented at strategically selected intersections across three states in Malaysia: Terengganu, Kedah, and Selangor. These locations were chosen to evaluate the system's performance under diverse traffic conditions and urban contexts. The locations are as follows:

1. **Terengganu:** The STL was deployed along a 6.8 km corridor on Jalan Sultan Mahmud, extending towards Jalan Kuantan–Kuala Terengganu in Kuala Terengganu. This corridor encompasses five interconnected intersections (J1 to J5). The site was purposively selected due to its high traffic volumes, particularly during peak hours, and its strategic importance in connecting key areas of the city. The interconnected nature of the intersections provided a valuable opportunity to assess the system's ability to coordinate traffic flow across multiple junctions.
2. **Kedah:** The STL was implemented along a 7.3 km corridor on Jalan Lencong Timur, encompassing five intersections (J1 to J5). This corridor was chosen due to its consis-

tently high traffic density and significant congestion, particularly during commuting hours. It serves as a vital artery for both local and regional traffic, making it an ideal location to evaluate the system's capacity to manage and optimize high-volume traffic flow.

3. **Selangor:** The STL was installed along a 3.7 km corridor on Jalan Subang, extending towards Persiaran Jubli Perak. This included eight intersections (PE26 to PE44). This corridor was selected for its significant traffic congestion, particularly during peak commuting hours, and its importance as a major arterial road connecting to key highways within the Selangor city center. The recurring congestion provided a suitable setting to evaluate the system's ability to manage and optimize traffic flow on a major urban thoroughfare. Figures 2 and 3 below are the study location of STL.

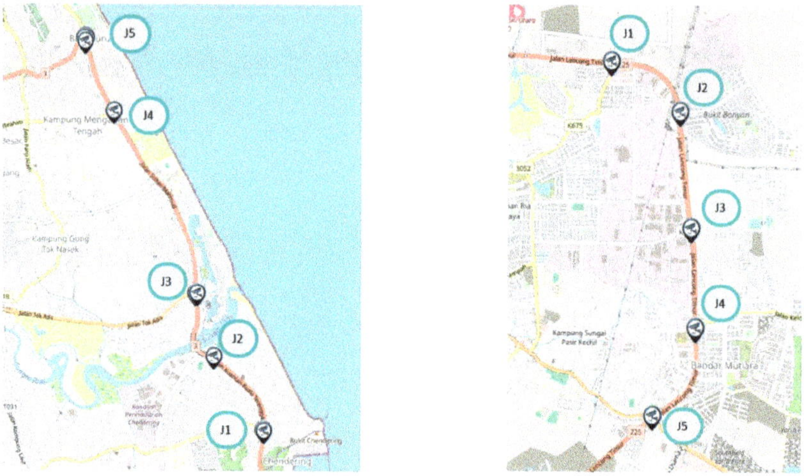

Figure 2. Location of intersection that implemented STL in Terengganu (**left**) and Kedah (**right**).

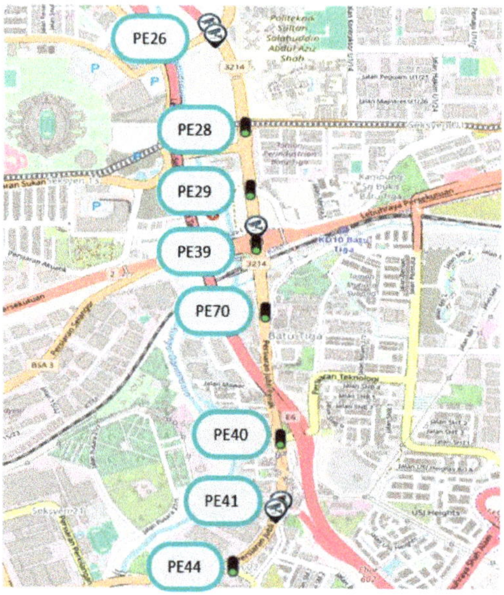

Figure 3. Location of intersection that implemented STL in Selangor.

3.3. Data Analysis and System Optimization

The data collected was used to dynamically adjust traffic light timings, prioritizing traffic flow based on real-time conditions. The AI algorithms at the heart of the STL were trained using historical traffic data from each location. These algorithms were further refined through continuous machine learning, enhancing their predictive accuracy and responsiveness to changing traffic patterns. This iterative learning process allows the system to move beyond pre-programmed timing sequences, adapting intelligently to optimize traffic flow in a way that traditional traffic lights cannot. This methodology was designed to specifically evaluate the advanced capabilities of the STL, demonstrating how its intelligent features enable it to surpass the limitations of traditional traffic lights towards dynamic and responsive traffic management. The results of this implementation and analysis will be presented in the following section.

4. Result and Findings

This section presents the results of the STL's implementation, demonstrating its capabilities beyond the limitations of traditional traffic signal control. The analysis focuses on the key performance indicator of travel time, showcasing the system's ability to dynamically optimize traffic flow.

Travel Time Analysis

The analysis of travel time data collected from Terengganu, Kedah, and Selangor sites revealed compelling evidence supporting the positive impact of the STL on traffic flow. Preliminary findings showcased a statistically significant reduction in traveling times across all locations and various data collection times, as illustrated in Table 1, Figures 4–9.

Table 1. Result of STL implementation in Terengganu, Kedah and Selangor.

Location	Data Collection Time	Average Travel Time Before STL (s)	Average Travel Time After STL (s)	Percentage Improvement (%)
Terengganu	(7.00 am–8.30 am)	422.7	392.6	7.10
Terengganu	(5.00 pm–6.30 pm)	445.5	400.5	10.10
Kedah	(7.00 am–8.30 am)	738.9	523.9	28.60
Kedah	(5.00 pm–6.30 pm)	720.7	597.3	16.40
Selangor	(7.00 am–8.30 am)	662.9	540.3	18.4
Selangor	(5.00 pm–6.30 pm)	790.3	635.6	18.8

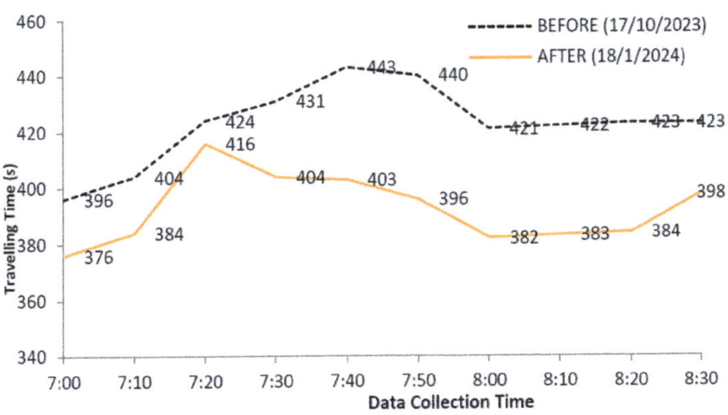

Figure 4. The traveling time in Terengganu STL location (7.00 am–8.30 am).

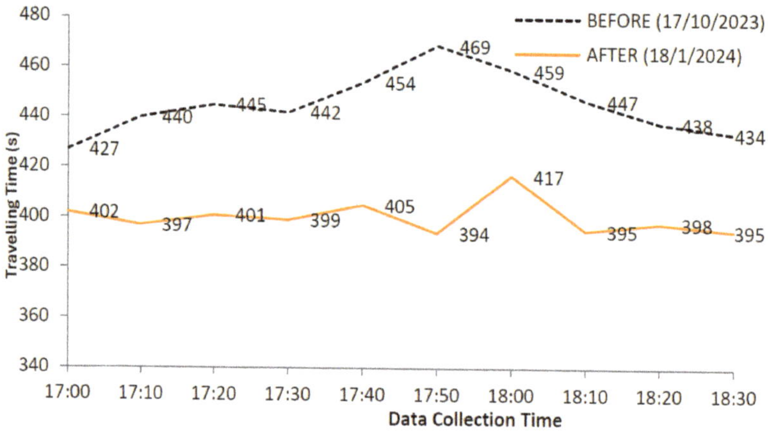

Figure 5. The traveling time in Terengganu STL location (5.00 pm–6.30 pm).

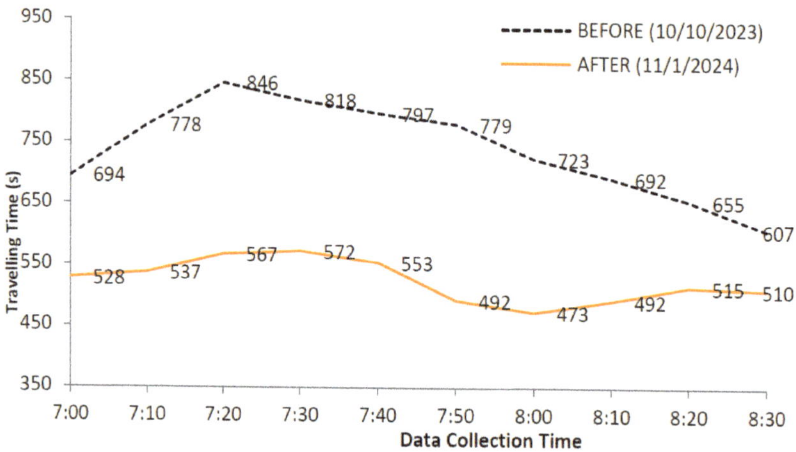

Figure 6. The traveling time in Kedah STL location (7.00 am–8.30 am).

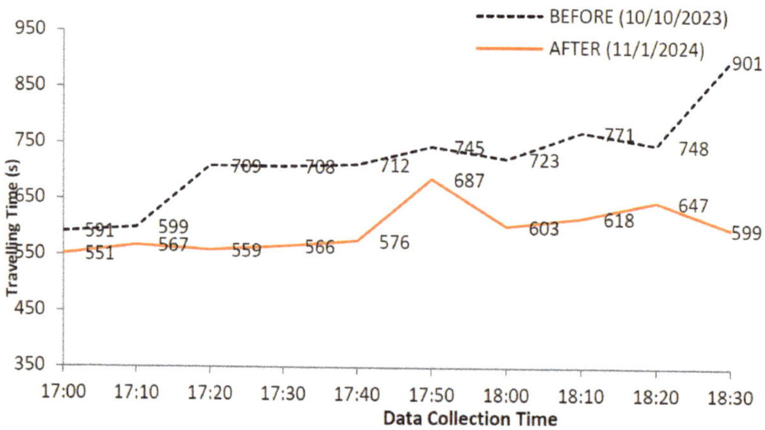

Figure 7. The traveling time in Kedah STL location (5.00 pm–6.30 pm).

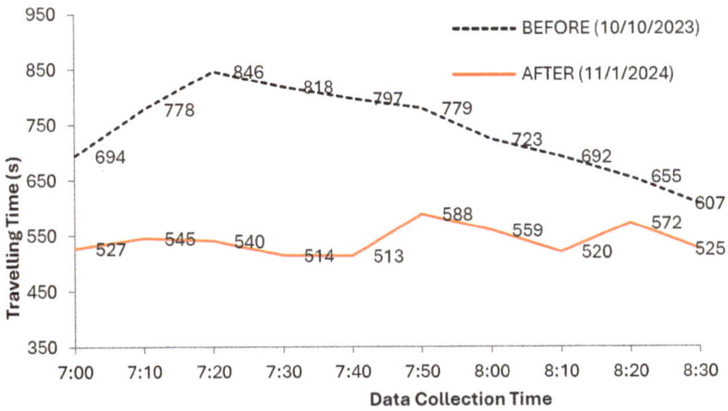

Figure 8. The traveling time in Selangor STL location (7.00 am–8.30 am).

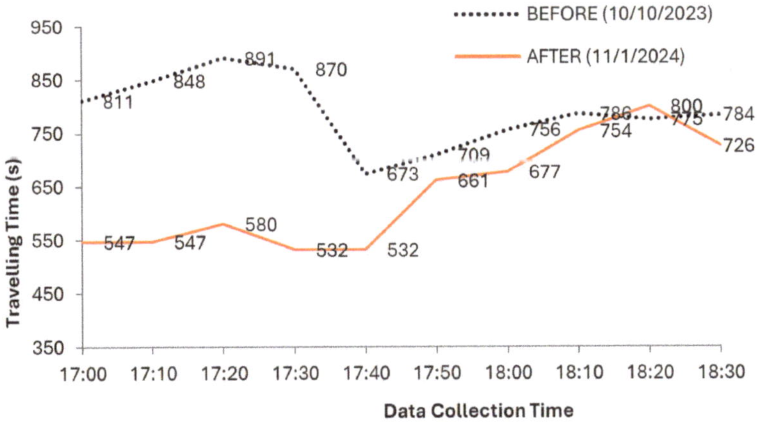

Figure 9. The traveling time in Selangor STL location (5.00 pm–6.30 pm).

These results underscore the efficacy of the STL in adapting to real-time traffic conditions and optimizing signal timings to reduce delays and enhance traffic flow efficiency. Moving beyond the fixed-time cycles of conventional traffic lights, the STL dynamically adjusts signal phases based on actual traffic demand, as evidenced by the observed reductions in travel time. The variations in improvement percentages between locations and times of day further highlight the system's ability to respond to the unique traffic patterns and demands of different areas and periods. The substantial reductions in travel times observed in all states—Terengganu, Kedah and Selangor—translate to significant time savings for commuters, leading to improved productivity, reduced fuel consumption, and lower emissions. These benefits have positive implications not only for individuals but also for businesses and the overall economy. Moreover, the improved traffic flow efficiency facilitated by the STL can contribute to a decrease in traffic congestion, further enhancing the quality of life in urban areas. This demonstrates the potential of intelligent traffic management systems to create more sustainable and livable cities, a clear advancement beyond the capabilities of traditional traffic signal technology. While this paper focuses primarily on travel time analysis, future research will delve into other performance indicators, such as queue length and stop time. This subsequent phase of the study will provide a more comprehensive understanding of the STL's overall impact on traffic network performance and further highlight its advanced capabilities beyond the red and green.

5. Conclusions

This study has demonstrated the effectiveness of STL in mitigating urban traffic congestion, showcasing their capabilities that extend far beyond the limitations of traditional red and green signal systems. By integrating AI, ANPR, and real-time data analysis, STL dynamically optimizes traffic light timings, leading to significant improvements in travel times and traffic flow. The system's adaptability to varying traffic conditions, as evidenced by the results from implementations in the states of Terengganu, Kedah and Selangor, is central to these achievements. This research underscores that the true potential of STL technology lies in its ability to learn and respond to real-time traffic dynamics, a clear departure from the static nature of conventional traffic signals. Beyond enhancing traffic efficiency, STL shows promise in improving road safety and environmental sustainability by reducing fuel consumption and emissions and further research is needed to explore long-term implications, scalability, and cost-effectiveness. The implementation of STL technology aligns with the Malaysian government's focus on AI and ITS to address urban challenges, particularly within the framework of the National Intelligent Transportation Management Center (NITMC) initiative. As Malaysia advances towards becoming a high-tech nation, adopting innovative solutions like STL will be crucial in creating sustainable, efficient, and resilient transportation networks. This study highlights the transformative potential of STL technology, illustrating how it is not merely controlling traffic but actively shaping a new era of urban mobility. By moving beyond the simple red and green, smart traffic lights are paving the way for a brighter, more connected future, where intelligent systems optimize traffic flow, enhance safety, and contribute to a better quality of life.

Author Contributions: Conceptualization, M.F.M., M.N.M., L.C.S., A.I.H.H. and A.M.H.; methodology, M.F.M., M.N.M., L.C.S. and A.I.H.H.; software, M.F.M., M.N.M., L.C.S. and A.I.H.H.; validation, M.F.M., M.N.M., L.C.S., A.I.H.H. and A.M.H.; formal analysis, M.F.M., M.N.M., L.C.S. and A.I.H.H.; investigation, M.F.M., M.N.M., L.C.S. and A.I.H.H.; resources, M.F.M., M.N.M., L.C.S. and A.I.H.H.; data curation, M.F.M., M.N.M., L.C.S., A.I.H.H. and A.M.H.; writing—original draft preparation, M.F.M., M.N.M. and A.I.H.H.; writing—review and editing, M.F.M., M.N.M., L.C.S., A.I.H.H. and A.M.H.; visualization, M.F.M., M.N.M., L.C.S., A.I.H.H. and A.M.H.; supervision, M.F.M. and M.N.M.; project administration, M.F.M., M.N.M., L.C.S. and A.I.H.H.; funding acquisition, M.F.M., M.N.M., L.C.S. and A.I.H.H. All authors have read and agreed to the published version of the manuscript.

Funding: This Smart Traffic Light project is funded by Government of Malaysia under The 12th Malaysia Plan (2021–2025) national budget.

Institutional Review Board Statement: Not applicable.

Informed Consent Statement: Not applicable.

Data Availability Statement: All the data are described in the paper.

Acknowledgments: The authors express their sincere gratitude to the following individuals and organizations for their invaluable support of this research: The members of the project team from the Highway Planning Division of the Ministry of Works Malaysia, the Malaysia Public Works Department (Electrical Branch HQ), and the Terengganu, Kedah, and Selangor Public Works Departments (Electrical Branch); their collaboration was instrumental in the successful completion of this project.

Conflicts of Interest: The authors declare no conflict of interest.

References

1. Sims, A.G.; Dobinson, K.W. The Sydney Coordinated Adaptive Traffic (SCAT) System Philosophy and Benefits. *IEEE Trans. Veh. Technol.* **1980**, *29*, 130–137. [CrossRef]
2. Guler, S.I.; Menendez, M.; Meier, L. Using connected vehicle technology to improve the efficiency of intersections. *Transp. Res. Part C Emerg. Technol.* **2014**, *46*, 121–131. [CrossRef]

3. Lee, W.H.; Chiu, C.Y. Design and Implementation of a Smart Traffic Signal Control System for Smart City Applications. *Sensors* **2020**, *20*, 508. [CrossRef] [PubMed]
4. Qin, X.; Khan, A.M. Control strategies of traffic signal timing transition for emergency vehicle pre-emption. *Transp. Res. Part C Emerg. Technol.* **2012**, *25*, 1–17. [CrossRef]
5. Ministry of Works Malaysia. *Malaysian ITS Blueprint 2019–2023*; Ministry of Works Malaysia: Kuala Lumpur, Malaysia, 2019.
6. Ministry of Transport Malaysia. *National Transport Policy 2019–2030*; Ministry of Transport Malaysia: Putrajaya, Malaysia, 2019.
7. Dewan Bandaraya Kuala Lumpur. Kuala Lumpur Smart City Master Plan 2021–2025. Available online: https://www.dbkl.gov.my/penerbitan-dan-laporan/kuala-lumpur-smart-city-blue-print-2021-2025 (accessed on 16 July 2025).
8. MIROS. *The Guideline on Implementing Tidal Flow Lane System*, 1st ed.; Malaysian Institute of Road Safety Research: Kajang, Malaysia, 2023.
9. Various Initiatives Prove Government's Seriousness in Driving Ai Development-Analyst (Bernama). Available online: https://www.bernama.com/en/news.php?id=2296694 (accessed on 11 May 2024).
10. Malaysia National AI Office. About NAIO. Available online: https://ai.gov.my/about-naio (accessed on 1 May 2024).

engineering proceedings

MDPI

Proceeding Paper

Pedestrian Model Development and Optimization for Subway Station Users †

Geon Hee Kim and Jooyong Lee *

Department of Urban and Transportation Engineering, Kyonggi University, Suwon Campus, Suwon-si 16227, Gyeonggi-do, Republic of Korea; 20241101142@kyonggi.ac.kr
* Correspondence: jy_lee@kyonggi.ac.kr
† Presented at the 2025 Suwon ITS Asia Pacific Forum, Suwon, Republic of Korea, 28–30 May 2025.

Abstract

This study presents an AI-enhanced pedestrian simulation model for subway stations, combining the Social Force Model (SFM) with LiDAR trajectory data from Samseong Station in Seoul. To reflect time-dependent behavioral differences, RMSProp-based optimization is performed separately for the morning peak, leisure hours, and evening peak, yielding time-specific parameter sets. Compared to baseline models with static parameters, the proposed method reduces prediction errors (MSE) by 50.1% to 84.7%. The model integrates adaptive learning rates, mini-batch training, and L2 regularization, enabling robust convergence and generalization across varied pedestrian densities. Its accuracy and modular design support real-world applications such as pre-construction design testing, post-opening monitoring, and capacity planning. The framework also contributes to Sustainable Urban Mobility Plans (SUMPs) by enabling predictive, data-driven evaluation of pedestrian flow dynamics in complex station environments.

Keywords: pedestrian traffic flow; social force model; rmsprop optimization; pedestrian trajectory analysis; crowd simulation

1. Introduction

In recent years, subways have gained renewed attention as a key component of sustainable urban transportation systems due to their environmental benefits and operational reliability. Enhancing the service quality of subway systems requires a balanced approach that encompasses both network-level expansions—such as new lines and transfer hub—and improvements to the physical environment within subway stations themselves. However, most prior research has predominantly focused on the former, addressing line extensions and intermodal connectivity, while studies examining the internal dynamics of pedestrian flow and operational efficiency within subway stations remain relatively scarce.

This research gap is increasingly problematic in the context of current urban transportation trends, which emphasize the complexification and upscaling of transit nodes. Modern subway stations are evolving into multi-functional complexes that integrate commercial, business, and cultural spaces, forming highly intricate pedestrian environments. A prominent example is the study site of this paper, Samseong Station in Seoul, which is currently undergoing development into a large-scale intermodal transit center scheduled for completion in 2028 [1]. Once completed, the station will feature a vertically and horizontally layered spatial structure that connects underground and aboveground facilities. These structural transformations are expected to generate non-linear and highly variable pedestrian flow patterns, necessitating precise modeling and monitoring capabilities to ensure operational safety and efficiency [2,3].

Moreover, as urban populations age and demand for inclusive mobility increases, ensuring safe and comfortable access for transportation-disadvantaged groups, such as older adults and people with disabilities, has become a critical objective. Key infrastructure components—such as elevators, ramps, and navigational aids—must be continuously monitored and maintained. This requires a high-fidelity pedestrian flow model capable of accurately reproducing real-world movement patterns and dynamic crowd interactions within complex station environments [4].

In sustainability-oriented transportation research, evaluation criteria must extend beyond environmental considerations such as greenhouse gas emissions or air pollutant reduction. A comprehensive framework must also account for pedestrian safety, inclusive accessibility, temporal demand variability, and adaptive service quality [5]. Pedestrian demand in subway stations exhibits substantial temporal variation. During the morning and evening peaks (07:00–09:00 and 18:00–19:00), stations—particularly those adjacent to major office and commercial complexes—tend to experience high levels of crowding and congestion due to concentrated commuting activity. In contrast, during midday leisure hours (14:00–14:30), which represent off-peak periods, pedestrian volumes are relatively lower, and individual walking behaviors become more pronounced, often reflecting personal spatial preferences rather than collective movement patterns.

These temporal differences highlight the importance of modeling pedestrian flow separately by time period, as pedestrian patterns vary significantly across different hours of the day [6]. Conducting simulations based on distinct time-of-day scenarios allows for more accurate prediction of crowd dynamics and individual walking behavior [7]. Moreover, incorporating external factors—such as weather conditions, local events, and infrastructure malfunctions—can further enhance the realism of the model and improve its applicability for planning and operational policy development.

To address these challenges, this study employs the Social Force Model (SFM) to develop a calibrated, data-driven simulation of pedestrian flow within Samseong Station. By leveraging real-world pedestrian trajectory data, the study conducts time-of-day-specific parameter optimization to reproduce observed patterns of interaction and congestion. Through this process, the model aims to evaluate and improve pedestrian safety, walking comfort, and operational resilience within complex subway station environments. The objective of this study aligns with the promotion of sustainable urban transport strategies in accordance with the guidelines of Sustainable Urban Mobility Plans (SUMPs), particularly in the context of the growing adoption of Intelligent Info-mobility Systems. Ultimately, this research contributes to the development of sustainable, inclusive, and intelligently managed urban transportation systems. Unlike conventional studies that focus primarily on transit network expansions, this work emphasizes in-station pedestrian dynamics as a vital dimension of sustainability, offering practical insights into efficient facility operation and user-centered design in large-scale, multifunctional transit hubs.

2. Literature Review

2.1. Social Force Model: Structure and Extensions

Originally proposed by Helbing and Molnár (1995), the Social Force Model (SFM) remains a foundational framework in pedestrian dynamics research [8]. The model describes pedestrian motion as the result of three core forces: a driving force, which reflects the pedestrian's intention to reach a destination at a desired speed; a repulsive force, which accounts for collision avoidance with other pedestrians and obstacles; and an attractive force, which models tendencies to approach companions, shops, or other points of interest.

Owing to its intuitive formulation and interpretability grounded in physical analogies, the SFM has been widely adopted in both academic studies and commercial simulation platforms such as VISSIM.

However, the original SFM adopts a deterministic approach, applying uniform parameter values across all pedestrians regardless of individual differences. As a result, the model fails to account for heterogeneity in physical capacity, psychological preferences, or contextual behavior, limiting its predictive accuracy in real-world environments. To address this limitation, Han et al. (2022) proposed an extended version of the SFM that incorporates fuzzy inference, enabling the model to simulate individual decision-making under uncertainty [7]. This enhancement allows the model to capture more realistic interpersonal interactions by incorporating linguistic variables and fuzzy logic into the force-based framework.

Recent advancements have further expanded the scope of SFM by applying it to shared spaces involving vehicle–pedestrian interactions. In particular, Yang et al. (2020) introduced a force component framework that differentiates interaction effects based on direction (e.g., front, rear, lateral) [9]. Their model, calibrated and validated using empirical trajectory data, demonstrated credible performance in replicating various types of interactions. However, it relied on the assumption of homogeneous parameters across all pedestrians, which limited its ability to accurately represent individual-level behavioral responses to vehicle interactions.

These findings underscore the importance of adapting model parameters based on contextual factors such as time-of-day congestion levels or travel purposes. Yang et al. recommended several improvements, including individual parameter calibration, customized loss functions, and interaction-type-specific formulations. These directions highlight the growing recognition that pedestrians should no longer be treated as reactive particles governed by uniform rules, but rather as adaptive agents interacting with heterogeneous and dynamic environments.

Consequently, advancing the SFM requires embedding behavioral heterogeneity, contextual sensitivity, and task-specific optimization strategies into the modeling process. Such enhancements will improve the realism, generalizability, and applicability of pedestrian simulations in complex urban scenarios.

2.2. Data-Driven Approaches: Machine Learning and Deep Learning

While rule-based pedestrian models such as the Social Force Model (SFM) provide interpretability and a clear physical foundation, they are limited in their ability to capture the full complexity and variability of pedestrian behavior—particularly in dynamic and non-standard environments. To address these limitations, data-driven approaches leveraging machine learning (ML) and deep learning (DL) have emerged as powerful alternatives capable of learning pedestrian patterns directly from trajectory data.

In the realm of machine learning, a wide range of supervised algorithms has been applied to enhance the realism of pedestrian simulations. Decision Trees and Random Forests have also been utilized to infer decision-making logic in complex environments [10]. These models typically convert features such as speed, direction, and local density into structured inputs for learning, and they often demonstrate high predictive accuracy even with relatively small datasets.

Deep learning further advances pedestrian modeling by enabling the capture of high-dimensional, nonlinear relationships in large-scale data. Convolutional Neural Networks (CNNs) are frequently employed to extract spatial features and predict pedestrian flow fields [11], whereas Long Short-Term Memory (LSTM) networks excel in learning temporal patterns and forecasting future trajectories [12]. Generative Adversarial Networks (GANs) have also been used to augment trajectory datasets by synthesizing realistic movements [13], and Reinforcement Learning (RL) has been applied to simulate adaptive decision-making under dynamic conditions, such as emergency evacuations [14].

Collectively, ML and DL approaches offer an expanded modeling capacity by supporting adaptive, probabilistic, and individualized representations of pedestrian behavior. However, these approaches come with trade-offs. DL-based models often require substantial computational resources and large annotated datasets. Therefore, balancing model fidelity with deployment efficiency remains a key concern when applying such methods in real-world contexts.

Given the deterministic limitations of the classical SFM and its sensitivity to uniform parameter settings, the integration of ML and DL into pedestrian modeling is not only promising but necessary. These data-driven methods can complement the interpretable structure of physics-based models like SFM, enhancing their behavioral realism and predictive accuracy. In complex urban environments where pedestrian movement is influenced by diverse contextual factors, a hybrid framework that combines SFM with ML/DL-based components—such as parameter calibration, trajectory prediction, or decision modeling—can provide a more flexible and scalable solution for pedestrian simulation.

3. Analysis of Pedestrian Behavior in Subway Stations

3.1. Study Area

Samseong Station, located at the intersection of Teheran-ro and Yeongdong-daero in Gangnam-gu, Seoul, is a major subway hub characterized by a multi-level structure that integrates both underground and above-ground facilities. Its strong connectivity to surrounding commercial zones, including COEX, has made it a critical node for pedestrian activity in the city's transit network. The focus area of this study is the passenger concourse adjacent to the turnstiles leading to Seolleung-bound trains—recognized as a high-density pedestrian zone due to its proximity to transfer points and commercial accessways. As such, this area is expected to exhibit diverse and dynamic pedestrian behaviors.

According to public smart card transaction data released by the Seoul Open Data Plaza in 2024, Samseong Station ranks as the seventh most heavily used subway station in South Korea among 338 stations analyzed [15]. With over 100,000 daily boarding and alighting events, the station consistently handles substantial pedestrian volumes. Given the presence of extensive underground corridors that connect the station directly to COEX and other major commercial complexes, the actual pedestrian flow through the underground space is likely even higher. These conditions make Samseong Station particularly suitable for studying time-sensitive and spatially complex pedestrian dynamics in an urban transit setting

To further contextualize the study, pedestrian traffic on Wednesday, 12 July 2017—a typical weekday during the high-travel summer vacation period—was analyzed using public data from the Seoul Open Data Portal. As shown in Table 1, the station recorded 64,847 boardings and 67,206 alightings on that day. During the morning peak (07:00–09:00), 90.4% of trips were alightings, reflecting intense inbound commuter flows. In contrast, during the midday leisure hours (14:00–15:00), boarding and alighting were nearly balanced, suggesting movement patterns shaped by shopping and tourism. In the evening peak (17:00–19:00), 78.5% of transactions were boardings, indicating high outbound demand.

Table 1. Pedestrian traffic at Samseong Station on 12 July 2017.

Time	**Morning Peak (07:00–09:00)**		**Leasure Hours (14:00–15:00)**		**Evening Peak (17:00–19:00)**	
Direction	Boarding	Alighting	Boarding	Alighting	Boarding	Alighting
Pedestrian traffic	2171	21,545	2630	2759	19,761	6168

These temporal asymmetries in boarding–alighting patterns emphasize the importance of time-sensitive pedestrian management strategies, particularly near major exits. The observed midday movement toward COEX also underscores the potential for coordination between transit infrastructure and adjacent commercial entities. By incorporating such time-dependent behavioral variations into simulation models, this study aims to improve the predictive accuracy and operational relevance of pedestrian flow modeling in complex subway environments.

3.2. LiDAR Data Collection and Preprocessing

In combination with the passenger flow data, this paper also utilizes pedestrian trajectory data collected via LiDAR (Light Detection and Ranging, Model: L-T1103-MRS57B, Manufacturer: SICK Ltd., Waldkirch, Germany) sensors on the same day (12 July 2017). The pedestrian trajectory data used in this study were collected inside a subway station in Seoul, as part of a prior study by Jo et al. (2018), which aimed to analyze pedestrian trajectory patterns based on LiDAR-based pedestrian tracking [16]. LiDAR technology enables high-resolution capture of pedestrian positions, velocities, accelerations, and movement directions. LiDAR data were collected near the turnstiles leading to Seolleung-bound trains. Figure 1a illustrates the trajectory data collected via LiDAR, while Figure 1b presents the detailed station layout of Samseong Station. The dataset includes Unix time, trajectory ID, x, y, and z coordinates, velocity, acceleration, direction, and angular velocity. Kalman filtering was applied to refine trajectory data and correct occlusions near the flap gate.

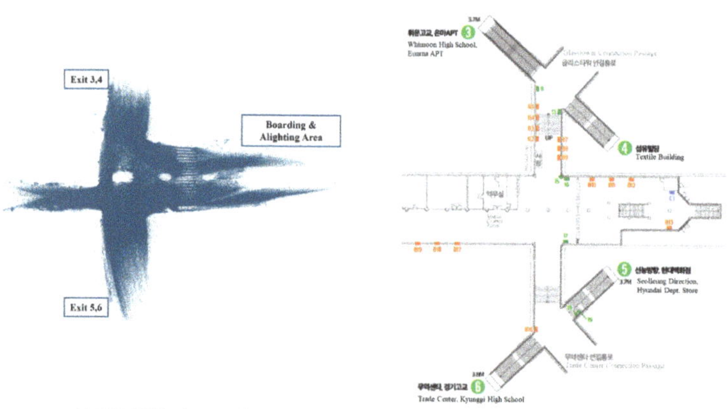

(**a**) LiDAR Trajectory Data (**b**) Samseong Station Layout

Figure 1. Trajectory data collected via LiDAR and the station layout of Samseong Station.

In Figure 1b, the numbers within green circles indicate the subway exits at Samseong Station, and the gray-colored text represents various facilities located within the underground station area.

From the LiDAR trajectory data, the characteristics of subway passengers at Samseong Station can be analyzed in relation to their connectivity with surrounding facilities. Pedestrians choosing specific exits tend to reflect the characteristics of the destinations connected to those exits. For example, exits 3 and 4 of Samseong Station are directly connected to the COEX shopping complex, suggesting that passengers using these exits are likely traveling for shopping and leisure activities. In contrast, exits 5 and 6 are connected to major office buildings, indicating that passengers using these exits are predominantly commuting to and from work.

Trajectory analysis revealed that during the morning peak hours, approximately 51% of pedestrian flow was directed toward exits 3 and 4. During leisure hours, pedestrian movement to exits 3 and 4 accounted for only 22%. In the evening peak hours, approximately 25% of passengers used exits 3 and 4. These findings highlight the correlation between exit selection and travel purpose, demonstrating how subway passenger behavior is influenced by the surrounding environment.

The fundamental flow–density relationship was used to analyze pedestrian traffic flow with the equation $q = u{\cdot}k$, where Q represents the pedestrian flow rate (person/min/m), u is the average pedestrian velocity (m/s), and k is the pedestrian density (person/m^2). The pedestrian density was estimated based on a total walkable area of 133 m^2, while velocity values were derived from LiDAR trajectory data. Figure 2 illustrates the flow–density relationship.

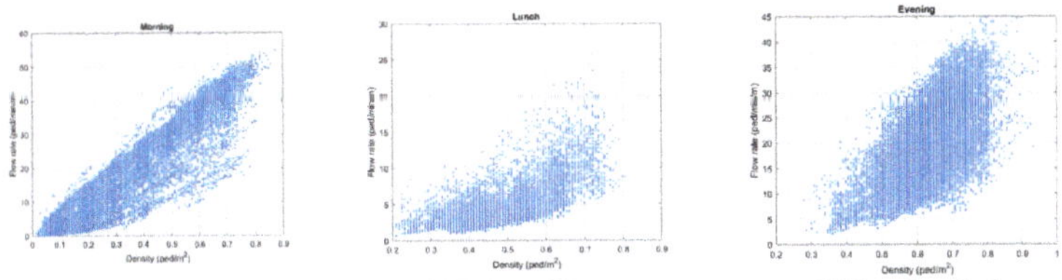

(**a**) Morning Peak (**b**) Leisure Hours (**c**) Evening Peak

Figure 2. Fundamental flow–density diagram for pedestrian traffic flow (by time of day).

The results show that during the morning peak hours (07:00–09:00), the Level of Service (LOS) ranged from C to D, indicating moderate to high congestion. During leisure hours (14:00–15:00), the LOS ranged from A to B, suggesting relatively smooth pedestrian movement. In the evening peak hours (17:00–19:00), the LOS ranged from B to C, highlighting increasing congestion due to high boarding rates.

4. Model Development

4.1. Mathematical Formulation of the Social Force Model

Pedestrian movement in complex environments is influenced by various factors, including individual intention, interactions with other pedestrians, and environmental constraints. The Social Force Model (SFM) provides a force-based approach to simulate these dynamics, capturing both self-propulsion and repulsive interactions [8]. By formulating pedestrian motion as a function of social forces, the model effectively represents realistic walking behaviors in high-density areas [17,18].

In this model, the total force f_i acting on pedestrian i is given by:

$$\vec{f_i} = m_i \frac{d\vec{v_i}}{dt} = m_i \left(\frac{\vec{v_i^0}(t) - \vec{v_i}(t)}{\tau_i} \right) + \sum_{j \neq i} \vec{f_{ij}} + \sum_{w} \vec{f_{iw}} \quad (1)$$

where:

- $\vec{f_i}$ denotes the total force acting on pedestrian i
- m_i is the pedestrian's mass

- $\vec{v_i^0}(t)$ is the current velocity at time t
- $\vec{v_i^0}(t)$ is the desired velocity at time t
- τ_i represents the relaxation time (i.e., how quickly a pedestrian adapts to the desired velocity)
- $\vec{f_{ij}}$ is the repulsive force from another pedestrian j
- $\vec{f_{iw}}$ is the repulsive force from a wall or obstacle W

The repulsive interaction force with other pedestrians is modeled as:

$$\vec{f_{ij}} = \left\{ A_i exp\left(\frac{r_{ij} - d_{ij}}{B_i}\right)\vec{n}_{ij} + \varphi g(r_{ij} - d_{ij})\vec{n}_{ij} \right\} n_{ij} + \omega g(r_{ij} - d_{ij})\Delta \vec{v}^{t}_{ji}\vec{t}_{ij} \quad (2)$$

Similarly, the repulsive interaction with walls or obstacles, modeled as a psychological avoidance behavior, is given by:

$$\vec{f_{iw}} = \left\{ A_i exp\left(\frac{r_i - d_{iw}}{B_i}\right)\vec{n}_{iw} + \varphi g(r_i - d_{iw})\vec{n}_{iw} \right\}\vec{n}_{iw} + \omega g(r_i - d_{iw})\left(\vec{v}_i \cdot \vec{t}_{iw}\right)\vec{t}_{iw} \quad (3)$$

where:

- d_{ij} and d_{iw} are distances from pedestrian i to pedestrian j or wall W, respectively
- $r_{ij} = r_i + r_j$ the sum of their radii
- $\vec{n}_{ij}$, $\vec{n}_{iw}$ are unit normal vectors (pointing from j or W toward i)
- $\vec{t}_{ij}$, $\vec{t}_{iw}$ are tangential direction vectors
- $\vec{\Delta v_{ji}} = \left(\vec{v}_j - \vec{v}_i\right) \cdot \vec{t}_{ij}$, the tangential component of the relative velocity
- $g(x)$ is a ramp function defined as:

$$g(x) = x \ if \ x > 0$$

This formulation captures both intentional and reactive components of pedestrian behavior, making the SFM particularly suitable for simulating complex environments such as subway stations. To enhance model fidelity, six parameters are calibrated using empirical trajectory data: reaction time (τ_i), desired velocity ($\vec{v_i^0}(t)$), social force constant (A_i), distance scale (B_i), body force (φ), and sliding friction force (ω). A dedicated optimization strategy is introduced in the following section to estimate these parameters effectively based on real-world trajectory data.

Since the Social Force Model generates different pedestrian trajectories depending on the parameter settings, accurate calibration of these parameters is essential for achieving high simulation accuracy [19]. Improper parameter configurations can lead to unrealistic behaviors, such as unnatural acceleration or collision patterns, especially in dense and dynamic environments. Therefore, the optimization process plays a critical role in ensuring that the model faithfully replicates the nuanced variations in pedestrian flow observed in actual subway station contexts.

4.2. RMSProp-Based Optimization Strategy

To calibrate the SFM parameters, this study proposes an integrated optimization framework that combines adaptive learning rates (RMSProp), mini-batch training, and L2 regularization.

RMSProp dynamically adjusts the learning rate for each parameter based on the variance of recent gradients, making it particularly well-suited for problems involving a large number of parameters, as is the case in the current study [20]. The algorithm provides stable convergence in regions with dense pedestrian interactions while enabling rapid optimization in sparse data environments, making it an effective calibration method for large-scale pedestrian dynamics simulations [17].

Unlike conventional optimization techniques such as Genetic Algorithms (GA) or Differential Evolution (DE), which are often computationally intensive and slow to converge, RMSProp has been widely recognized in deep learning for its efficiency and robustness. Leveraging these strengths, the algorithm is used in this study to calibrate the key parameters of the Social Force Model (SFM) using empirical pedestrian trajectory data.

To further improve optimization efficiency and generalization performance, several complementary strategies are incorporated, including mini-batch training, L2 regularization, and early stopping. Mini-batch training allows for stable and scalable handling of large trajectory datasets, while L2 regularization suppresses excessive parameter magnitudes and mitigates overfitting. In addition, the early stopping technique automatically terminates training when the validation loss fails to improve over a fixed number of iterations, reducing redundant computation and enhancing the model's generalizability.

The loss function is defined as the mean squared error (MSE) between the predicted and observed pedestrian velocity vectors, serving as a quantitative measure of how accurately the model replicates real-world movement patterns. The complete RMSProp-based optimization procedure is summarized as a pseudo-code in Table 2.

Table 2. Pseudo-code for RMSProp-based parameter optimization of the Social Force Model.

Input: α (RMSProp decay rate), γ (learning rate), θ_0 (initial parameter set), *LOSS* (mean squared error), **Initialize:** $v_o \leftarrow 0$ (Set moving average of squared gradients to zero) *best_params* $\leftarrow \theta_0$ (Set the best parameters to the initial values) *best_loss* $\leftarrow \infty$ (Initialize the best loss to infinity)
Main Loop: For each mini-batch: Step 1. Predict velocities with Social Force Model: Step 2. Loss Calculation: $LOSS = \frac{1}{N}\sum_{i=1}^{N}(v_i - \hat{v}_i)^2$ Step 3. Gradient Computation: $g_t = \nabla_\theta LOSS$ Step 4: RMSProp Gradient Update: $v_t = \alpha v_{t-1} + (1-\alpha)g_t^2$ Step 5: Parameter Update $\theta_t = \theta_{t-1} - \gamma\frac{g_t}{\sqrt{v_t + \epsilon}}$ Step 6: Update *best_params* $\leftarrow \theta_t$ and *best_loss* $\leftarrow$ *LOSS* if conditions met Step 7: Early Stopping Check If no improvements: Terminate
Output: Return optimal parameter set θ

4.3. Simulation Setup

The simulation environment was constructed to replicate the physical layout of the observed pedestrian space. Four rectangular exits were placed along the domain boundaries, corresponding to real-world egress points at Samseong Station. To represent physical constraints, two vertical and two horizontal static walls were added. Each pedestrian was assigned a mass of 80 kg, and the simulation time step (Δt) was set to 0.2 s.

Initial values for the six SFM parameters—reaction time (τ_i), desired speed (${v_i}^0$), social force constant (A_i), distance scale (B_i), body force (φ), and sliding friction force (ω)—were set based on prior literature (0.5, 1.3, 2000, 0.08, 1.2×10^5, 2.4×10^5) and randomly perturbed within predefined bounds informed by domain expertise. The RMSProp-based optimization was conducted separately for three time segments—morning peak (07:00–09:00), leisure hours (14:00–15:00), and evening peak (17:00–19:00)—using trajectory data collected via LiDAR sensors.

To ensure efficiency and generalization, a mini-batch size of 1024 was used. RMSProp parameters included a learning rate of 1×10^{-4}, decay rate (β) of 0.9, and $\varepsilon = 1 \times 10^{-8}$. Parameter updates were bounded within plausible ranges, and L2 regularization ($\lambda = 1 \times 10^{-4}$) was applied to prevent overfitting.

5. Simulation Results

5.1. Training and Validation Loss Analysis

Model performance was evaluated by computing the mean squared error (MSE) between predicted and observed pedestrian velocity vectors in both the x and y directions. The training and validation loss curves for each time segment are presented in Figure 3, illustrating stable convergence under varying pedestrian flow conditions. These results indicate that the RMSProp-based calibration effectively reduced prediction errors without overfitting, thereby supporting the reliability of the simulation results.

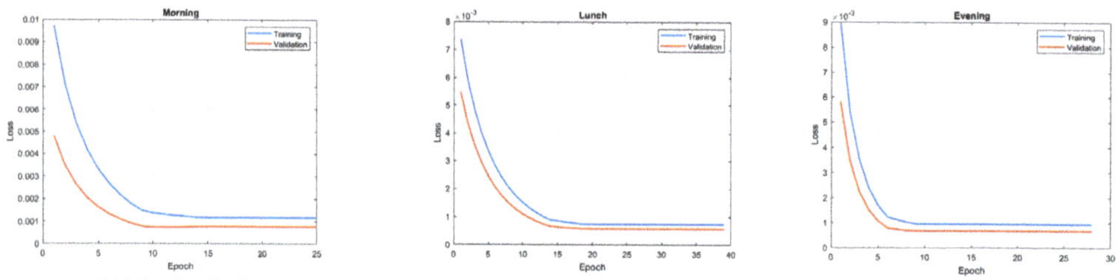

(**a**) Morning Peak (**b**) Leisure Hours (**c**) Evening Peak

Figure 3. Training validation curves of (**a**) morning peak, (**b**) leisure hours, (**c**) evening peak.

The optimization algorithm was applied to three datasets corresponding to distinct time periods: the morning peak (07:00–09:00), leisure hours (14:00–14:30), and the evening peak (18:00–19:00). Each period exhibited different pedestrian dynamics, suggesting the potential need for time-specific parameter optimization. Nevertheless, a uniform model structure was applied across all datasets. The learning curves revealed three consistent patterns:

i. rapid convergence in the early phase due to adaptive learning rate adjustment;
ii. stable decline in both training and validation losses during the mid phase;
iii. early stopping in the final phase to prevent overfitting.

Figure 3 illustrates these learning and validation curves for each time of day, highlighting the model's adaptability to various levels of crowd density.

5.2. Time-Specific SFM Parameter Analysis

The optimized parameters of the Social Force Model varied significantly by time period. During the morning peak, the desired velocity was relatively high at 0.66 m/s, reflecting the urgency of commuter behavior. In contrast, the leisure period exhibited lower interaction intensity and more balanced movement patterns. During the evening peak, the social force coefficients increased, indicating the need to capture complex pedestrian interactions and congestion patterns during that time. Table 3 presents the optimized parameter values for different time periods. The parameter optimization results revealed variations across different time periods. During the morning peak, commuters exhibited a higher desired velocity (0.66 m/s) due to their urgency to reach their destinations. In leisure hours, pedestrian interactions were relatively lower, resulting in more balanced movement patterns. In the evening peak, the social force parameters increased, reflecting complex pedestrian interactions and congestion management needs.

Table 3. Parameter optimization results by time periods.

Parameter	τ_i (s)	$v_i^0(t)$ (m/s)	A_i(N)	B_i(N)	φ (kg·s^{-2})	ω (kg·m^{-1}·s^{-1})
Morning Peak	1.00	0.66	1990.99	0.05	119,506.60	241,587.90
Leisure Hours	1.00	0.50	2069.95	0.05	126,146.60	226,296.80
Evening Peak	1.00	0.50	2170.85	0.05	117,868.50	245,919.70

5.3. Comparing MSE Between Before and After Optimization

The effectiveness of the optimization was further assessed by comparing MSE values before and after training. As illustrated in Figure 4, the MSE decreased from 0.1221 to 0.0609 during the morning peak, from 0.3738 to 0.0572 during leisure hours, and from 0.0965 to 0.0381 during the evening peak. This represents a reduction ranging from a minimum of 50.1% to a maximum of 84.7%, demonstrating an enhanced explanatory power in capturing pedestrian behavior. This result suggests that the MSE before optimization varies by time of day, indicating that the default SFM's performance depends on specific pedestrian flow characteristics. In contrast, the MSE after optimization suggests that similar performance can be achieved for each time-of-day dataset once the optimization is completed, thereby enhancing the model's reliability and credibility.

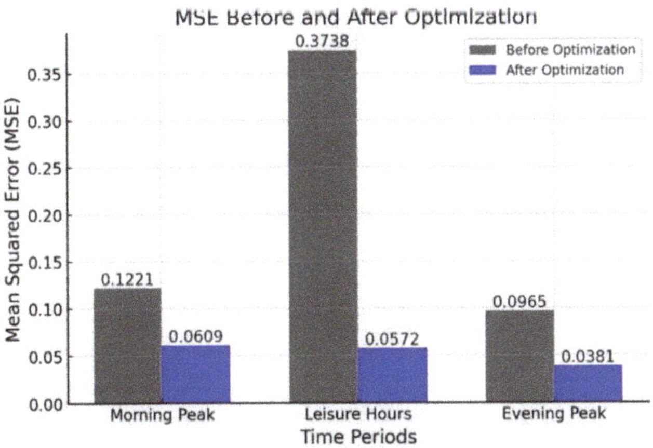

Figure 4. Reduction of mean squared error (MSE) After Optimization (by time-of-day).

The overall accuracy of the model's prediction performance improved after optimization, as further illustrated in Figure 5. A time-series comparative analysis revealed several key findings: During the morning peak, the optimized model generated predictions that closely aligned with the observed average speed, accurately capturing the initial low-speed walking state. During leisure hours, the predicted speed better reflected variations in observed speed and exhibited greater sensitivity to periodic fluctuations in walking velocity. In the evening peak, the pre-optimization model struggled to represent atypical walking patterns accurately.

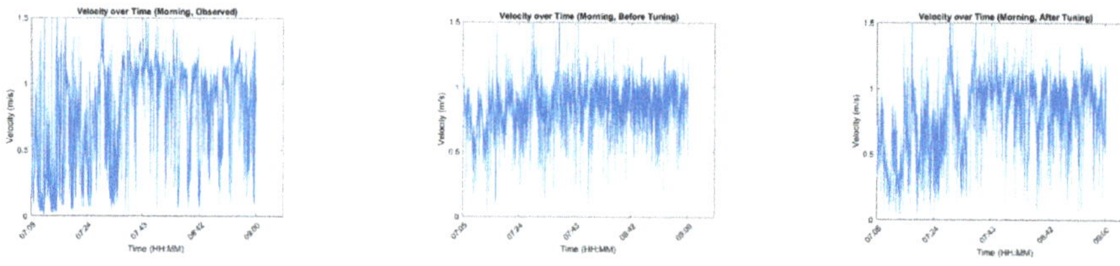

Figure 5. Velocity over time during morning peak: observed, before optimization, and after optimization.

5.4. Velocity Prediction Accuracy Analysis

To assess the temporal accuracy of the model, the predicted pedestrian speeds (both before and after optimization) were compared with the observed speeds for each time of day. The velocity-over-time plots for each period are presented in Figures 5–7, where each figure consists of three panels: the left panel shows the observed pedestrian speeds, the middle panel shows the speeds predicted by the before-optimization model, and the right panel shows the speeds predicted by the after-optimization model. Specifically, Figure 5 presents results for the morning peak and for the leisure hours, and Figure 7 presents results for the evening peak period.

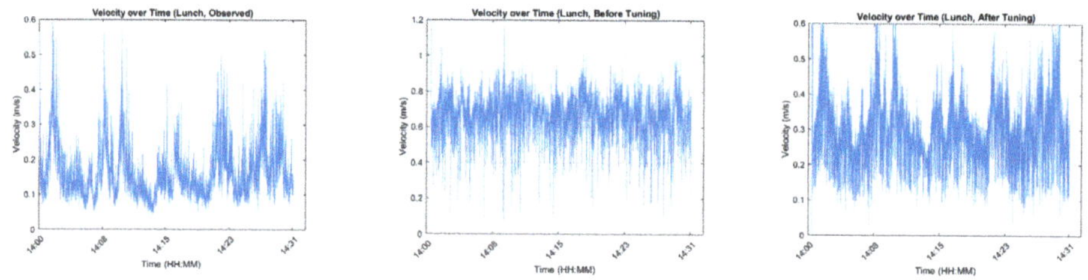

Figure 6. Velocity over time during leisure hours: observed, before optimization, and after optimization.

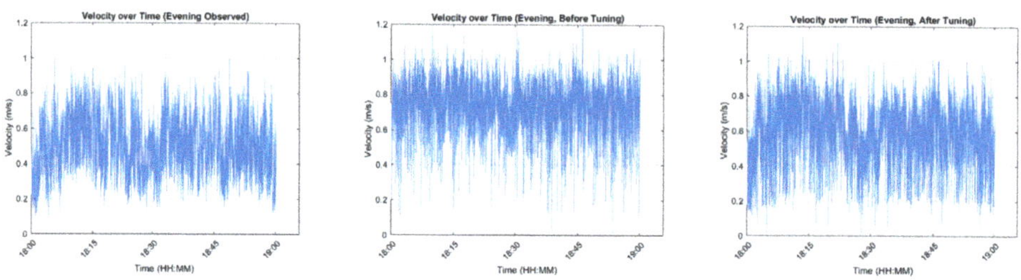

Figure 7. Velocity over time during evening peak: observed, before optimization, and after optimization.

During the morning peak, the optimized model more accurately captured the initial low-speed walking state observed in the empirical data. For the leisure period, the model effectively reproduced periodic fluctuations in walking speed. Notably, the observed trajectory data revealed a recurring pattern in which walking speeds increased at intervals of approximately 2–3 min. The optimized model successfully predicted this phenomenon, reflecting a realistic response to train arrival schedules in the subway station—an important feature of pedestrian dynamics during this period. In the evening peak, the pre-optimization model failed to capture irregular walking behaviors. However, after optimization, the model was able to reflect such atypical patterns more clearly. Visual inspection of the velocity curves confirms that the optimized model produced significantly more accurate predictions under complex evening conditions.

Overall, the before-optimization model exhibited noticeable errors in segments characterized by periodic speed fluctuations, whereas the after-optimization model substantially reduced these errors and accurately reproduced time-dependent walking dynamics across all time periods. These findings confirm that the Social Force Model can effectively capture pedestrian behavior in diverse subway station environments when its parameters are rigorously calibrated through an appropriate optimization process.

6. Conclusions

This study developed a pedestrian-flow simulation tailored to the intricacies of Samseong Station in Seoul by fusing high-resolution LiDAR trajectories with a Social Force Model (SFM) whose parameters were rigorously calibrated via an RMSProp optimization scheme. Segmenting the dataset into three distinct time-of-day intervals—morning peak, leisure hours, and evening peak—enabled the model to reproduce the pronounced temporal heterogeneity in walking behavior that arises from shifting trip purposes and fluctuating density levels.

The RMSProp-based calibration markedly enhanced predictive fidelity, most notably in velocity estimation. The optimized model faithfully replicated gradual acceleration patterns characteristic of the morning commute, cyclic speed oscillations during midday leisure activity, and the erratic dynamics associated with evening congestion. Crucially, it retained high reliability under dense, unstable flow conditions, underscoring its robustness for real-world deployment.

Beyond its empirical accuracy, the model offers immediate operational value for congestion mitigation and real-time station management, and its computational efficiency makes it well-suited for integration into digital twin platforms that support dynamic scenario analysis and emergency-response training. Ultimately, the proposed modeling approach serves as a foundational tool for the implementation of Sustainable Urban Mobility Plans (SUMPs), particularly as Intelligent Info-mobility Systems gain momentum in urban infrastructure. By aligning data-driven simulation with policy-oriented objectives, this research contributes to the creation of mobility strategies that are inclusive, adaptive, and sustainable.

Nevertheless, this study primarily focused on conventional pedestrian interactions. Future research should aim to incorporate more complex behavioral mechanisms, including responses to environmental cues such as signage, spatial constraints, and real-time congestion feedback. Furthermore, accounting for individual-level heterogeneity—such as age, baggage possession, and other personal attributes—along with group-based dynamics will be essential for enhancing the model's descriptive richness and applicability. The establishment of standardized benchmarking frameworks is also crucial for evaluating model performance across diverse subway station types and structural configurations. Addressing these research directions will substantially improve the predictive accuracy, scalability, and generalizability of pedestrian simulation models, thereby advancing the design of safer, more intelligent, and operationally efficient subway systems.

Author Contributions: Conceptualization, G.H.K. and J.L.; methodology, G.H.K.; software, G.H.K.; validation, G.H.K. and J.L.; formal analysis, G.H.K.; writing—original draft preparation, G.H.K.; writing—review and editing, J.L.; supervision, J.L. All authors have read and agreed to the published version of the manuscript.

Funding: This work was supported by the IITP (Institute of Information & Communications Technology Planning & Evaluation)-ICAN (ICT Challenge and Advanced Network of HRD) grant funded by the Korea government (Ministry of Science and ICT) (IITP-2025-RS-2024-00436954).

Institutional Review Board Statement: Not applicable.

Informed Consent Statement: Not applicable.

Data Availability Statement: Data is not available due to privacy issues of observed pedestrians.

Conflicts of Interest: The authors declare no conflict of interest.

References

1. Yeongdong-Daero Underground Space Complex Development. Available online: https://uri.seoul.go.kr/web/contents/57.do?mid=1149 (accessed on 10 April 2025).
2. Seok, Y.S.; Kang, C.H. A study on planning of multi-modal transfer centers in railway stations. *Proc. Archit. Inst. Korea Conf.* **2018**, *38*, 255–258. (In Korean)
3. Kim, J.H.; Han, S.Y.; Lee, U.D. A study on congestion levels in subway station environments. *Korean Soc. Railw. Conf.* **2008**, *6*, 1810–1816. (In Korean)
4. Wang, S.; Wang, Z. Collaborative development and transportation volume regulation strategy for an urban agglomeration. *Sustain.* **2023**, *15*, 14742. [CrossRef]
5. Zhao, H.T.; Yang, S.; Chen, X.X. Cellular automata model for urban road traffic flow considering pedestrian crossing street. *Phys. A Stat. Mech. Its Appl.* **2016**, *462*, 1301–1313. [CrossRef]
6. Calabrò, G. A new agent-based model to simulate demand-responsive transit in small-sized cities. *Sustainability* **2025**, *17*, 5279. [CrossRef]
7. Han, M.J. A pedestrian space analysis program reflecting pedestrian behavior. *Railw. J.* **2015**, *18*, 103–108. (In Korean)
8. Helbing, D.; Molnar, P. Social force model for pedestrian dynamics. *Phys. Rev. E* **1995**, *51*, 4282. [CrossRef]
9. Yang, D.; Özgüner, Ü.; Redmill, K. A social force based pedestrian motion model considering multi-pedestrian interaction with a vehicle. *ACM Trans. Spat. Algorithms Syst.* **2020**, *6*, 1–27. [CrossRef]
10. Shafaghat, A. Path walkability assessment framework based on decision tree analysis of pedestrian travelers' retail walking. Ph.D. Thesis, Universiti Teknologi Malaysia, Skudai, Malaysis, March 2013.
11. Zamboni, S.; Kefato, Z.T.; Girdzijauskas, S.; Norén, C.; Dal Col, L. Pedestrian trajectory prediction with convolutional neural networks. *Pattern Recognit.* **2022**, *121*, 108252. [CrossRef]
12. Xue, H.; Huynh, D.Q.; Reynolds, M. SS-LSTM: A hierarchical LSTM model for pedestrian trajectory prediction. In Proceedings of the IEEE Winter Conference on Applications of Computer Vision (WACV), Lake Tahoe, NV, USA, 12–15 March 2018; pp. 1186–1194.
13. Fang, F.; Zhang, P.; Zhou, B.; Qian, K.; Gan, Y. Atten-GAN: Pedestrian trajectory prediction with GAN based on attention mechanism. *Cogn. Comput.* **2022**, *14*, 2296–2305. [CrossRef]
14. Everett, M.; Chen, Y.F.; How, J.P. Collision avoidance in pedestrian-rich environments with deep reinforcement learning. *IEEE Access.* **2021**, *9*, 10357–10377. [CrossRef]
15. Seoul Metropolitan City. Transportation Policy Division. Available online: https://data.seoul.go.kr/dataList/OA-12914/S/1/datasetView.do (accessed on 10 April 2025).
16. Jo, Y.; Jeong, E.; You, S.I.; Oh, C. Trajectory pattern analysis using LiDAR-based pedestrian tracking. *J. Korean Soc. Transp.* **2018**, *36*, 503–518. (In Korean) [CrossRef]
17. Lee, J.; Kim, T.; Chung, J.H.; Kim, J. Modeling lane formation in pedestrian counter flow and its effect on capacity. *KSCE J. Civ. Eng.* **2016**, *20*, 1099–1108. [CrossRef]
18. Zeng, W.; Chen, P.; Nakamura, H.; Iryo-Asano, M. Application of social force model to pedestrian behavior analysis at signalized crosswalk. *Transp. Res. Part C Emerg. Technol.* **2014**, *40*, 143–159. [CrossRef]
19. Sticco, I.M.; Frank, G.A.; Dorso, C.O. Social force model parameter testing and optimization using a high stress real-life situation. *Phys. A: Stat. Mech. Its Appl.* **2021**, *561*, 125299. [CrossRef]
20. Zou, F.; Shen, L.; Jie, Z.; Zhang, W.; Liu, W. A sufficient condition for convergences of adam and rmsprop. In Proceedings of the IEEE/CVF Conference on Computer Vision and Pattern Recognition (CVPR), Long Beach, CA, USA, 15–20 June 2019; pp. 11119–11127.

engineering proceedings

MDPI

Proceeding Paper

Virtual Capacity Expansion of Stations in Bikesharing System: Potential Role of Single Station-Based Trips †

Gyugeun Yoon

Department of Civil & Environmental Engineering, Seoul National University, 1, Gwanak-ro, Gwanak-gu, Seoul 08826, Republic of Korea; gyugeun.yoon@snu.ac.kr; Tel.: +82-2-880-8394 or +82-2-873-2684

† Presented at the 2025 Suwon ITS Asia Pacific Forum, Suwon, Republic of Korea, 28–30 May 2025.

Abstract

Bikeshare systems usually relocate bikes to respond to a mismatch between demand and bike supply, imposing substantial costs to operators despite the effort to encourage users to participate in voluntary rebalancing. This study initiates a search for a new strategy that can involve single station-based (SSB) riders and consider their bikes as the reserve of the current bike balance, resulting in the virtual expansion of station capacity. Thus, the behaviors of bike riders related to SSB trips are compared to investigate the potential applications. The results from analyzing the data of Citi Bike in New York City indicate that 13.4% of total trips were SSB, and the average trips per origin and destination (OD) pair was 2.6 times higher. Also, distinctive characteristics such as mean trip time regarding user groups and bike types were statistically significant within numerous OD pairs, implying the need for separate policies for both groups. Based on the analysis, stations with the highest expected benefit are identified.

Keywords: bikeshare system; rebalancing policy; single station-based trips; user behavior; station capacity

1. Introduction

Bikeshare systems (BSSs) have been iconic urban mobility services in many global cities since they were introduced as an alternative to transportation modes depending on internal combustion engine vehicles. Meanwhile, it is natural that users of a BSS fall into two heterogenous groups: residents who conduct daily travels, and visitors who temporarily stay. We can guess their behavioral differences would be substantial due to the distinct circumstances that lead them to use bikeshares. One interesting point is the proportion of users who check out and return bikes at the same station. It is expected that short-term customers will be the majority of them, as they are more likely to ride bikes to look around nearby. These "single station-based (SSB)" trips can play an important role in BSS station capacity management, functioning as an additional storage of bikes. This study, therefore, demonstrates data analyses that can verify the benefit of considering SSBs as one of the alternatives to manage dock capacity. Annual members and casual users are separately grouped to take into account their behavioral differences.

2. Relevant Literature and Research Gap

Bikeshare users choose between annual membership and a casual pass according to attributes such as their length of stay, trip purposes, or "willingness-to-cycle", affecting discrepancies in trip behaviors. Thus, many studies have shown the heterogeneity in

BSS user groups and distinguished riders by the passes they held. While several studies included them as factors influencing the result [1], some addressed direct comparisons of two user types as the main research objective to derive meaningful insights [2,3].

BSS station capacity, usually equivalent to the number of docks, affects the convenience of users [4]. However, excessive capacity may lead to an increasing need for rebalancing bikes since the station retains more bikes to be relocated to other stations unless it is one of the popular pickup points. Therefore, planners designate the location and specification of bike stations by expected demand distribution at the planning stage [5]. Although a rebalancing strategy based on portable stations that can function as independent stations or be appended to existing ones was proposed [6], operators need to pay attention to the relocation of portable stations and become reluctant to actively apply capacity management to their system.

Instead of modifying the number of docks, it may be possible to "virtually" keep bikes at stations by incorporating bike fleets for SSB trips as the "reserve". Stations with many SSB trips can have the potential to attract more demand by providing larger bike fleets if the number of bikes to be returned is appropriately predicted. In this context, this study can provide a background for the development of a BSS station capacity management scheme that ensures simple implementation and modification without physical adjustment.

3. Methodology

3.1. Preprocessing of Citi Bike Historical Usage Dataset

Monthly usage data with abnormal records and daily data associated with the precipitation history from NOAA Online Weather Data (NOWData) were deleted. There remained four bike-user groups since there are two pass types and two bike types: members with classic bikes (MC), members with e-bikes (ME), casual users with classic bikes (CC), and casual users with e-bikes (CE). First, pass type is the only clue that can differentiate the customer type between casual users and annual members, assuming that they relate with them. Second, different bike types affect the trip duration, one of the core factors compared in this study.

3.2. Single Station-Based Trip Identification

The objective of this part is to verify if the composition of bike-user types in SSB trips is differentiated from that in trips between different stations regarding both the amount and proportion of trips. Among several possible reasons to return to the point where they checked bikes out, riding a bike as a recreational activity such as physical exercise or jaunts may also be a primary purpose. Then, it can be expected that casual users may perform this kind of trip pattern more frequently than annual members. Although members would take more recreational trips in total, the proportion may be higher with casual users. The contingency table approach can be adequately applied to this question [7].

4. Results and Discussion

Table 1 shows the composition of trips. Only OD pairs with more than six trips were chosen to conduct the two-sample Kolmogorov–Smirnov (K-S) test, a non-parametric statistical test based on empirically cumulated density function, which requires at least six samples [8]. Figure 1 indicates the rejection rate of tests per paired type in different trip patterns.

The higher value means a larger discrepancy. First, trip duration distributions of classic and electric bike trips by casual users become similar if their trip pattern is an SSB OD pair. Second, for users who rode the same bike type, their trip duration distributions were more similar if their trips were not SSB. This result implies that casual users have higher chances to perceive riding the bike itself as a recreational activity and are less attentive to reducing their trip duration. From the results, annual members and casual users seem to show distinct bike usage behaviors in terms of spatiotemporal distribution and potential purpose of bike usage. This heterogeneity should be considered when designing a voluntary bike rebalancing program for casual users which has never been implemented in reality.

Table 1. Number of trips per trip types with non-zero trips.

Trip Type	Annual Members		Casual Users		Total
	Classic Bikes	Electric Bikes	Classic Bikes	Electric Bikes	
SSB trips (1306 pairs)	224,239 (43.7%)	109,164 (21.3%)	128,934 (25.1%)	50,429 (9.8%)	512,766 (100.0%)
Avg. trips per OD pair	171.7	83.6	98.7	38.6	392.6
Normal trips (21,622 pairs)	1,902,012 (57.5%)	551,973 (16.7%)	617,423 (18.7%)	238,598 (7.2%)	3,310,006 (100.0%)
Avg. trips per OD pair	88.0	25.5	28.6	11.0	153.1
Total	2,126,251	661,137	746,357	289,027	3,822,772

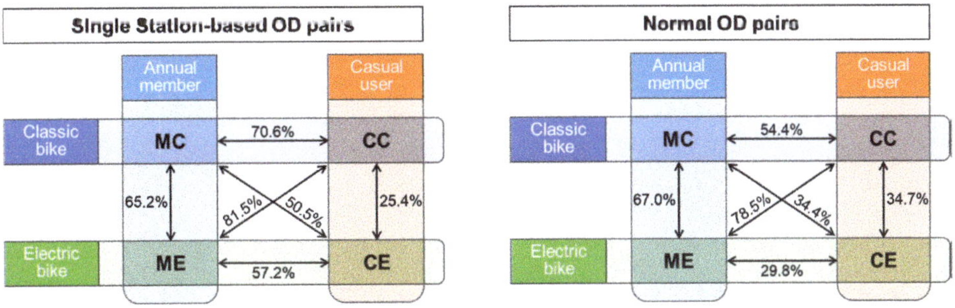

Figure 1. Rejection rate of K-S tests among bike-user groups for different trip types.

The most promising benefit of the proposed approach is clear: capacity expansion without additional massive capital investment and rebalancing effort. The system can induce SSB riders to bring their bikes back to the stations where they checked them out. A future research direction can aim to quantify the impact of the proposed virtual station capacity management. The system can add a function that asks whether a user would return the bike to the origin station. Collecting this information in advance can support a more sophisticated estimation of the number of available bikes. Moreover, additional surveys may reveal the relationship between compensation and participation rate, which should differ by user group.

Funding: This research received no external funding.

Institutional Review Board Statement: Not applicable.

Informed Consent Statement: Not applicable.

Data Availability Statement: The original data presented in the study are openly available on Citi Bike website: https://citibikenyc.com/system-data (accessed on 21 July 2025).

Conflicts of Interest: The author declares no conflict of interest.

References

1. Crossa, A.; Reilly, K.H.; Wang, S.M.; Lim, S.; Noyes, P. If we build it, who will come? Comparing sociodemographic characteristics of bike share subscribers, cyclists, and residents of New York City. *Transp. Res. Rec.* **2022**, *2676*, 634–642. [CrossRef]
2. Wergin, J.; Buehler, R. Where do bikeshare bikes actually go?: Analysis of capital bikeshare trips with GPS data. *Transp. Res. Rec.* **2017**, *2662*, 12–21. [CrossRef]
3. El-Assi, W.; Mahmoud, M.S.; Habib, K.N. Effects of built environment and weather on bike sharing demand: A station level analysis of commercial bike sharing in Toronto. *Transportation* **2017**, *44*, 589–613. [CrossRef]
4. Wang, K.; Chen, Y.J. Joint analysis of the impacts of built environment on bikeshare station capacity and trip attractions. *J. Transp. Geogr.* **2020**, *82*, 102603. [CrossRef]
5. Frade, I.; Ribeiro, A. Bike-sharing stations: A maximal covering location approach. *Transp. Res. A Policy Pract.* **2015**, *82*, 216–227. [CrossRef]
6. Almannaa, M.H.; Elhenawy, M.; Masoud, M.; Rakha, H.A. A New Mathematical Approach to Solve Bike Share System Station Imbalances Based On Portable Stations. In Proceedings of the 2019 IEEE Intelligent Transportation Systems Conference (ITSC 2019), Auckland, New Zealand, 27–30 October 2019; pp. 1721–1726.
7. National Institute of Standards and Technology. NIST/SEMATECH e-Handbook of Statistical Methods. 2012. Available online: https://www.itl.nist.gov/div898/handbook/prc/section4/prc46.htm (accessed on 10 February 2025).
8. Minitab. Two Sample Kolmogorov-Smirnov Normality Test of the Underlying Distributions—Minitab. Available online: https://support.minitab.com/en-us/minitab/18/macro-library/macro-files/nonparametrics-macros/kstwo/ (accessed on 10 February 2025).

engineering proceedings

MDPI

Proceeding Paper

Developing a Risk Recognition System Based on a Large Language Model for Autonomous Driving

Donggyu Min [1] and Dong-Kyu Kim [2,*]

[1] Department of Civil and Environmental Engineering, Seoul National University, Seoul 08826, Republic of Korea; dgmin@snu.ac.kr
[2] Department of Civil and Environmental Engineering and Institute of Construction and Environmental Engineering, Seoul National University, Seoul 08826, Republic of Korea
* Correspondence: dongkyukim@snu.ac.kr; Tel.: +82-28807348
† Presented at the 2025 Suwon ITS Asia Pacific Forum, Suwon, Republic of Korea, 28–30 May 2025.

Abstract

Autonomous driving systems have the potential to reduce traffic accidents dramatically; however, conventional modules often struggle to accurately detect risks in complex environments. This study presents a novel risk recognition system that integrates the reasoning capabilities of a large language model (LLM), specifically GPT-4, with traffic engineering domain knowledge. By incorporating surrogate safety measures such as time-to-collision (TTC) alongside traditional sensor and image data, our approach enhances the vehicle's ability to interpret and react to potentially dangerous situations. Utilizing the realistic 3D simulation environment of CARLA, the proposed framework extracts comprehensive data—including object identification, distance, TTC, and vehicle dynamics—and reformulates this information into natural language inputs for GPT-4. The LLM then provides risk assessments with detailed justifications, guiding the autonomous vehicle to execute appropriate control commands. The experimental results demonstrate that the LLM-based module outperforms conventional systems by maintaining safer distances, achieving more stable TTC values, and delivering smoother acceleration control during dangerous scenarios. This fusion of LLM reasoning with traffic engineering principles not only improves the reliability of risk recognition but also lays a robust foundation for future real-time applications and dataset development in autonomous driving safety.

Keywords: risk recognition system; autonomous driving; autonomous driving

1. Introduction

Recent NHTSA reports indicate that 94% of traffic accidents stem from human error [1], prompting significant interest in the use of autonomous driving for accident prevention [2]. However, current systems still struggle to interpret the complex interactions among small vehicles, emergency vehicles, and pedestrians [3]. To address these shortcomings, recent research has begun leveraging large language models (LLMs) such as GPT-4 [4], which exhibit emergent reasoning capabilities comparable to human judgment [5]. However, many studies overlook transportation domain knowledge—particularly surrogate safety measures (SSMs)—and rely on unrealistic 2D simulations.

This study enhances risk recognition and system reliability by integrating GPT-4 with the time-to-collision (TTC) metric in a realistic 3D simulation environment, CARLA 0.9.14. Our approach combines advanced LLM reasoning with critical transportation engineering insights, offering a promising direction for more robust and context-aware autonomous driving systems.

2. Methods

This study aims to detect potentially dangerous situations that are apparent to humans but challenging for conventional models. We implemented a dangerous scenario in CARLA, an open-source 3D simulation platform. Specifically, we selected the viaduct area of CARLA, where shadows and parked vehicles create visibility obstacles, as shown in Figure 1.

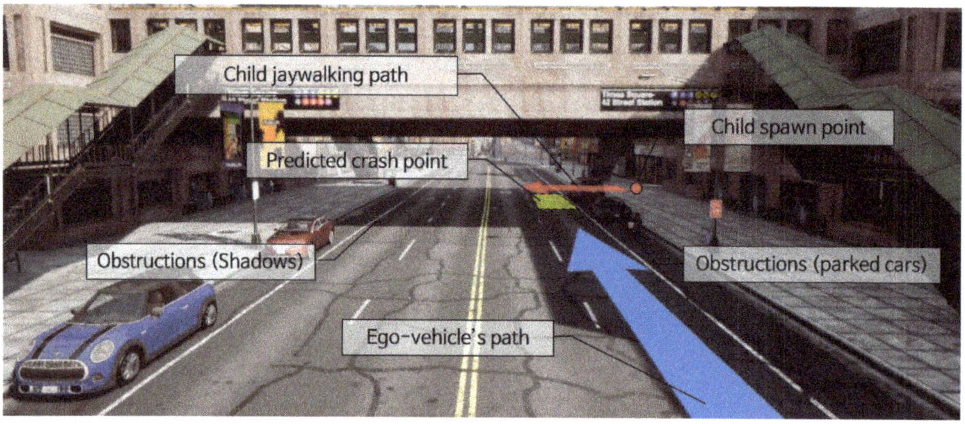

Figure 1. Experimental scenario with potential risk.

In our scenario, a child unexpectedly crosses in front of a parked vehicle at 5 m/s, requiring the autonomous vehicle to detect this risk from approximately 30 m away and respond appropriately. Our framework in Figure 2 operates the ego-vehicle in autonomous mode within CARLA and extracts key data: a front RGB image, four pieces of surrounding object information (object ID, type, Euclidean distance, and time-to-collision), and five driving parameters (speed, acceleration, throttle, steering, and brake). This information is reformatted into natural language and input into GPT-4, which evaluates the traffic situation and determines if a dangerous condition exists—responding with "YES" or "NO", along with its reasoning. Based on GPT-4's judgment, vehicle control commands are issued, and CARLA simulates the resulting traffic safety outcomes.

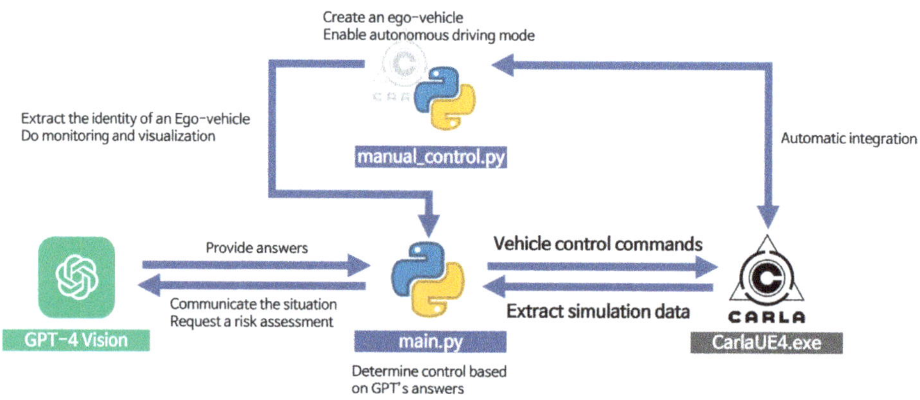

Figure 2. Modular roles of the proposed framework.

3. Results

As a result of the provided information and the request that GPT determine the risk situation, it was recommended that brakes be applied to the autonomous vehicle, as shown in Table 1. To evaluate performance, we extracted the distance to the child pedestrian, TTC, and vehicle acceleration at 0.1 s intervals. Figure 3 indicates that the LLM-based module significantly improves traffic safety compared to the default module. With the default module, the distance was reduced to 6.2 m at 1.7 s after control, whereas the LLM-based module maintained a distance of 11.8 m at 1.3 s, about 5.6 m greater. Moreover, the LLM-based module provided a more stable TTC, achieving 0.38 s at 1.2 s versus 0.04 s at 1.7 s with the default module. Additionally, the default module exhibited erratic acceleration patterns, failing to recognize dangerous situations. These findings highlight the benefits of integrating LLM with domain knowledge using TTC as a critical safety metric to enhance autonomous driving performance. While CARLA-based simulations validate the approach, future work must incorporate real-time capabilities. The proposed framework serves as a foundation for building a learning dataset for further advancements in autonomous perception technology.

Table 1. Results of GPT-4-based risk assessment.

GPT-4 Questions	GPT-4 Answers
The attached image is from the front camera of an autonomous vehicle, namely ego-vehicle. The following is the result of detecting the surrounding objects of ego-vehicle. The unit of Distance is "m" and the unit of TTC (Time to Collision) is "s". [Detected Objects] Object ID: 27, Type: walker.pedestrian.0049, Distance: 19.19, TTC: 2.0083130971714924 The following parameters represent the driving status of the Ego-vehicle. [Ego-vehicle Information] Speed: 9.69 m/s, Acceleration: (−2.40, −0.43, 0.00) m/s^2 Throttle: 0.00, Steering: −0.02, Brake: 0.15 Do not just consider numbers like the position of objects, but pay attention to the behavior of the vehicle or pedestrian in the image. In particular, be aware of potentially dangerous situations, such as jaywalking pedestrians. Even if the likelihood of a collision is low right now, you never know what might happen. Be prepared for the risk of a potential accident. Even a pedestrian on a pedestrian road can distinguish potential hazards from his gestures. Again, consider the context of the image as a whole. Given the three pieces of information above, does this ego-vehicle need to maintain or start the brakes? Answer YES or NO. When answering, answer yes or no first, and then take a paragraph to explain your reasoning. Be aware of visibility restrictions caused by parked cars. Pedestrians, including children, may appear suddenly. Detect the head, torso, feet, etc. of an obscured child within a limited field of view. If you see children, they might be playing and run into the driveway	Yes. Given the information, the ego-vehicle needs to maintain or, if necessary, apply more braking. The provided object detection shows that there is a pedestrian(walker.pedestrian.0049) at a distance of 19.19 m and with a Time to Collision (TTC) of approximately 2 s. Considering the speed of the ego-vehicle is 9.69 m/s, and the pedestrian is in close proximity to the vehicle's path, active braking is crucial to avoid a potential collision. Furthermore, the ego-vehicle's current brake value is 0.15, and its acceleration is negative, indicating that it is already decelerating, which should be continued or increased if needed, to assure the safety of the pedestrian and comply with traffic safety regulations. The image provided corroborates the need for caution. The pedestrian is visible on the sidewalk, and there are also parked cars that could obstruct the view of other potential pedestrians, especially children, who may enter the street unpredictably. Situational awareness is key in such environments, and the autonomous vehicle's systems must ac-count for obscured areas and the possibility of sudden pedestrian movements. The vehicle's de-cision algorithms should prioritize pedestrian safety and prepare to stop if necessary, considering the vehicle's speed and the pedestrian's proximity to the potential path of the vehicle.

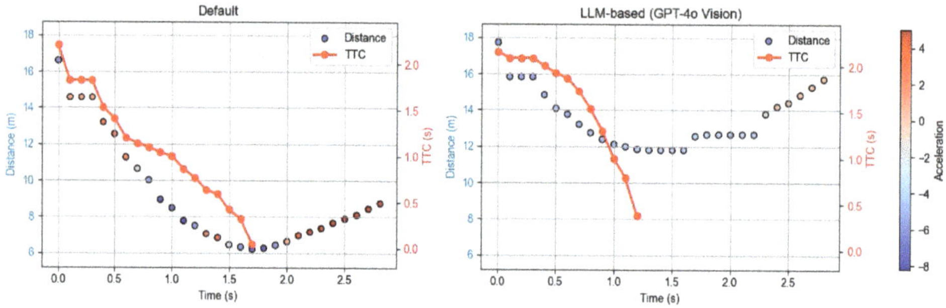

Figure 3. The safety performance of the default and LLM-based modules.

Author Contributions: Conceptualization, D.M. and D.-K.K.; methodology, D.M. and D.-K.K.; software, D.M.; validation, D.M.; formal analysis, D.M. and D.-K.K.; investigation, D.M.; resources, D.M.; data curation, D.M.; writing—original draft preparation, D.M.; writing—review and editing, D.M. and D.-K.K.; visualization, D.M.; supervision, D.-K.K.; project administration, D.M.; funding acquisition, D.-K.K. All authors have read and agreed to the published version of the manuscript.

Funding: This research is supported by the Korea Institute of Police Technology (No.092021C28S02000), the National Research Foundation of Korea (No.2022R1A2C2012835), and the Korea Ministry of Land, Infrastructure, and Transport's Innovative Talent Education Program for Smart City.

Institutional Review Board Statement: Not applicable.

Informed Consent Statement: Not applicable.

Data Availability Statement: The data presented in this study are available upon request from the corresponding author.

Conflicts of Interest: The authors declare no conflict of interest.

References

1. National Highway Traffic Safety Administration (NHTSA). Critical Reasons for Crashes Investigated in the National Motor Vehicle Crash Causation Survey. Available online: http://www-nrd.nhtsa.dot.gov/Pubs/812115.pdf (accessed on 15 February 2025).
2. Yurtsever, E.; Lambert, J.; Carballo, A.; Takeda, K. A survey of autonomous driving: Common practices and emerging technologies. *IEEE Access Digit. Object Identifier.* **2020**, *8*, 58443–58469. [CrossRef]
3. Fu, D.; Li, X.; Wen, L.; Dou, M.; Cai, P.; Shi, B.; Qiao, Y. Drive like a human: Rethinking autonomous driving with large language models. In Proceedings of the IEEE/CVF Winter Conference on Applications of Computer Vision, Waikoloa, HI, USA, 1–6 January 2024.
4. Achiam, J.; Adler, S.; Agarwal, S.; Ahmad, L.; Akkaya, I.; Aleman, F.; Almeida, D.; Altenschmidt, J.; Altman, S.; Anadkat, S.; et al. Gpt-4 technical report. *arXiv* **2023**, arXiv:2303.08774.
5. Wei, J.; Tay, Y.; Bommasani, R.; Raffel, C.; Zoph, B.; Borgeaud, S.; Yogatama, D.; Bosma, M.; Zhou, D.; Metzler, D.; et al. Emergent abilities of large language models. *arXiv* **2022**, arXiv:2206.07682.

engineering proceedings

MDPI

Proceeding Paper

Development of Detection and Prediction Response Technology for Black Ice Using Multi-Modal Imaging †

Seong-In Kang and Yoo-Seong Shin *

Korea Expressway Corporation Research Institute, Gimcheon-si 39660, Republic of Korea; sikang@ex.co.kr
* Correspondence: wato@ex.co.kr; Tel.: +82-031-8098-6358
† Presented at the 2025 Suwon ITS Asia Pacific Forum, Suwon, Republic of Korea, 28–30 May 2025.

Abstract

As traffic accidents caused by black ice during the winter continue to occur, there is a growing need for technologies that enable drivers to recognize and respond to black ice in advance. In particular, to reduce major accidents and associated casualties, it is essential to provide timely information and prevent incidents through accurate prediction. This paper proposes an artificial intelligence (AI) technology capable of detecting and predicting black ice using multimodal data. The study aims to enable a preemptive response in the field of digital disaster safety and discusses the applicability and effectiveness of the proposed approach in real-world road environments.

Keywords: expressway; black ice; artificial intelligent; prediction; multi-modal

1. Introduction

Black ice prediction technology typically relies on weather data. However, the commonly used data such as temperature, humidity, wind speed, rain, and snow is usually collected from highways and is often far away from the Expressway section being analyzed. Weather conditions can change rapidly in areas like shaded mountainous zones, and these environmental variations can significantly degrade prediction accuracy. Therefore, it is essential to obtain road-specific meteorological data and generate input data by calibrating the collected weather information according to the topography and road environment of the prediction location [1,2].

For early prediction, it is essential that artificial intelligence can recognize the initial signs of black ice formation. However, relying solely on basic meteorological data as input is insufficient for sensitively detecting the early changes on the road surface caused by black ice. Therefore, an AI model based on multimodal data that integrates various active data sources is required to improve both sensitivity and prediction accuracy [3,4].

As shown in Figure 1, technologies related to black ice detection and prediction have predominantly been developed and validated within simulated environments. Real-world applications remain largely limited to basic field deployments, resulting in low reliability of the derived outcomes. Consequently, the objective accuracy of these approaches remains uncertain, underscoring the need for systematic demonstration and validation to improve both technological maturity and predictive performance.

https://doi.org/10.3390/engproc2025102008

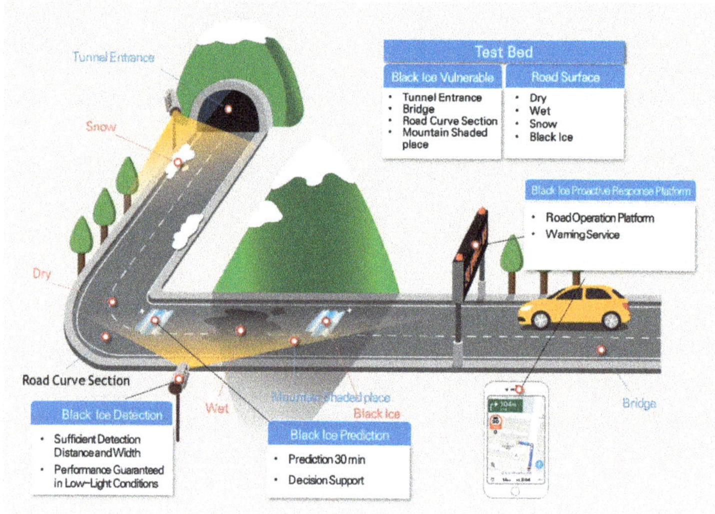

Figure 1. Black ice detection and prediction concept.

This study focuses on the development of an artificial intelligence technology capable of detecting and predicting black ice at an early stage using multimodal data, as well as the implementation and demonstration of a digital disaster safety platform for preemptive response.

2. Research Trends Related to Black Ice

As illustrated in Figure 2, a machine learning model utilizing the random forest algorithm was developed to estimate winter road surface conditions. This model leverages data collected from roadside weather sensors and CCTV video imagery. Additionally, a long short-term memory (LSTM)-based AI model was proposed to predict road surface temperatures at various time intervals by utilizing the time-series characteristics of meteorological data.

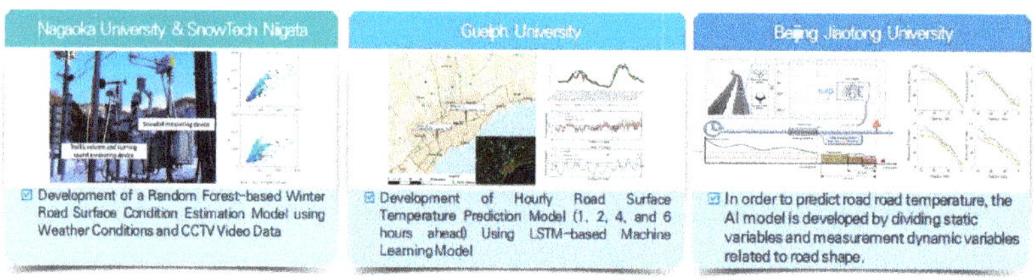

Figure 2. Black ice prediction.

The imaging sensor WIDE SENSE enables the acquisition of short-wave infrared (SWIR) spectral data, which can distinguish between ice and water layers on road surfaces, thus allowing accurate and continuous monitoring of road conditions (Italy).

Grip is a system designed to detect black ice by analyzing information collected from in-vehicle sensors and estimating the road surface friction coefficient in real time (Finland).

A prediction model for black ice occurrence on national expressways was developed by collecting weather data and road location information. Additionally, a black ice prediction model was developed by incorporating system dynamics concepts and utilizing black ice-related data together with road GIS information.

A road icing probability estimation model was developed by equipping regular patrol vehicles with temperature sensors to measure both ambient air and road surface temperatures, using the temperature difference as a key variable (KICT).

As illustrated in Figure 3, a system titled "Black Ice Detection Method and System Using the Same" was developed to improve detection reliability by implementing dual-sensor detection of black ice on road surfaces. This approach utilizes both LiDAR and thermal imaging cameras to enhance the accuracy and robustness of black ice identification.

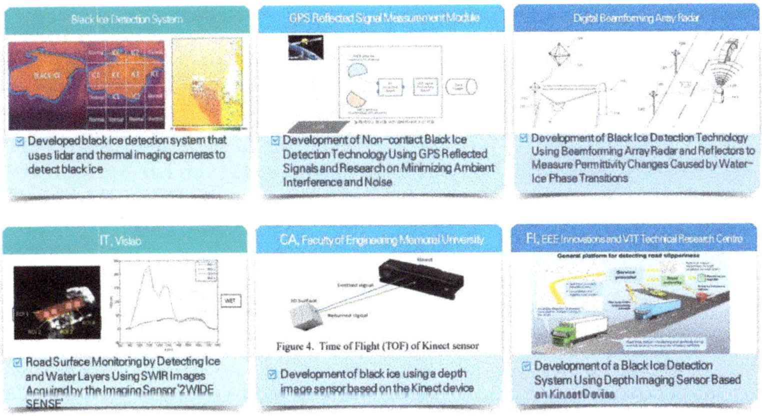

Figure 3. Black ice detection.

A black ice detection technique utilizing non-contact GPS signal reflections was developed, along with ongoing research aimed at minimizing the effects of environmental factors and signal noise.

Additionally, a black ice detection technology using beamforming array radar combined with reflectors was developed to measure variations in the dielectric constant caused by phase transitions between water and ice on road surfaces.

3. Contents of the Research

As shown in Figure 4, this research and development initiative aims to develop an artificial intelligence (AI)-based system for the early detection and prediction of black ice through the integration of multimodal data sources. These sources include meteorological sensors, CCTV video feeds, vehicle-mounted sensor data, and geographic information system (GIS) data. The fusion of these heterogeneous datasets facilitates more accurate and timely assessment of road surface conditions conducive to black ice formation.

To overcome the limitations of existing black ice detection systems—which are often confined to simulation environments or provide only generalized risk information—this project aims to develop a real-time, field-deployable solution. By employing machine learning techniques such as random forests, long short-term memory (LSTM) networks, and system dynamics modeling, the project seeks to construct robust prediction models that account for both temporal and spatial variability in weather and road conditions.

Furthermore, the project will implement a digital disaster safety platform designed to facilitate proactive responses by delivering tailored alerts and risk assessments to both road operators and drivers. This platform will support decision-making by providing real-time road status updates and risk forecasts up to 60 min in advance.

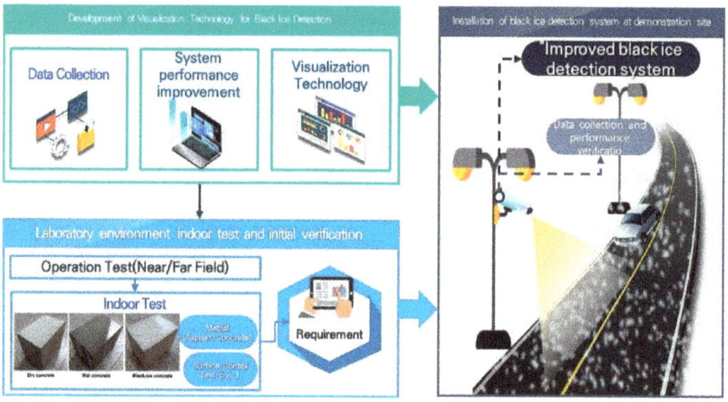

Figure 4. Black ice detection and prediction concept.

4. Test Bed Construction

Most technologies related to black ice detection have demonstrated accuracy only under ideal or laboratory conditions, and their deployment in real-world road environments has typically been limited to basic applications without comprehensive validation procedures.

Technologies that have not undergone proper field verification have reliability issues and are often utilized merely as reference materials. Consequently, these technologies cannot be effectively integrated into existing road operation platforms or utilized in practical response processes.

As shown in Figure 5, an experimental scenario simulating realistic driving conditions was constructed using test roads situated adjacent to operational expressways, such as the Yeo-ju Test Road in Korea. The scenario incorporated application modules specifically tailored to the varying pavement types present within the test area, thereby enhancing the representativeness and applicability of the evaluation environment.

Figure 5. Yeo-ju Test Road (Korea Expressway).

A total of 32 weather-related parameters are received every minute, including wind direction, wind speed, temperature, humidity, atmospheric pressure, rainfall detection, rainfall correction, road condition, and road surface temperature.

5. Conclusions

To ensure practical applicability and reliability, the proposed technologies will be tested and validated through field demonstrations in black ice-prone areas such as tunnel exits, shaded mountainous roads, curved sections, and intersections. The ultimate goal of this project is to enhance road safety by minimizing black ice-related accidents and enabling timely, data-driven interventions.

This research is expected to contribute to strengthening the competitiveness of domestic winter road management technologies by developing early black ice prediction techniques based on accurate detection within the target area and road surface condition forecasting.

Based on the predicted risk level and evaluation of the black ice prediction model's performance over different time intervals, optimal information and countermeasures can be proposed. This will support the development of AI-driven, decision-support technologies led by road managers.

Author Contributions: Project administration and writing—original draft preparation, S.-I.K.; writing—review and editing, Y.-S.S. All authors have read and agreed to the published version of the manuscript.

Funding: This work was supported by Korea Planning & Evaluation Institute of Industrial Technology funded by the Ministry of the Interior and Safety (MOIS, Korea) [Project Name: Development of Early Detection and Preemptive Response Technology for Black Ice Using Multi-modal Imaging; Project Number: RS-2024-00409314].

Institutional Review Board Statement: Not applicable.

Informed Consent Statement: Not applicable.

Data Availability Statement: Data are available in this manuscript.

Conflicts of Interest: The authors declare no conflict of interest.

References

1. Jeong, M.C.; Lee, J.H.; Oh, H.Y. Ensemble machine learning model-based YouTube spam comment detection. *J. Korea Inst. Inf. Commun. Eng.* **2020**, *24*, 576–583.
2. Kim, J.Y.; Lee, H.J.; Paik, J.R. Survey on distinction of black ice using sensors. *J. Korea Soc. Comput.* **2020**, *28*, 78–87.
3. Zheng, M.; Wu, S.; Wang, C.; Li, Y.; Ma, Z.; Peng, L. A Study on Evaluation and Application of Snowmelt Performance of Anti-Icing Asphalt Pavement. *Appl. Sci.* **2017**, *7*, 583. [CrossRef]
4. Park, G.Y. A case study on meteorological analysis of freezing rain and black ice formation on roads in winter. *J. Environ. Sci. Int.* **2017**, *267*, 827–836.

engineering proceedings

MDPI

Proceeding Paper

Development of a Multidirectional BLE Beacon-Based Radio-Positioning System for Vehicle Navigation in GNSS Shadow Roads †

Tae-Kyung Sung [1,*], Jae-Wook Kwon [2], Jun-Yeong Jang [2], Sung-Jin Kim [3] and Won-Woo Lee [4]

1 Department of Information Communications Engineering, Chungnam National University, 99 Daehak-ro, Yuseong-gu, Daejeon 34134, Republic of Korea
2 WiFive Ltd., 82 Daehak-ro, Yuseong-gu, Daejeon 34183, Republic of Korea; ck1015@wifive.co.kr (J.-W.K.); wkdwnsdud25@wifive.co.kr (J.-Y.J.)
3 Korea Automotive Technology Institute, 303 Pungse-ro, Pungse-myeon, Dongnam-gu, Cheonan 31214, Republic of Korea; sjkim@katech.re.kr
4 Korea Expressway Corporation Research Institute, 24 Dongtansunhwan-daero 17, Dongtan-myeon, Hwaseong 18489, Republic of Korea; wonwoo2.lee@gmail.com
* Correspondence: tksaint@cnu.ac.kr
† Presented at the 2025 Suwon ITS Asia Pacific Forum, Suwon, Republic of Korea, 28–30 May 2025.

Abstract

In outdoor environments, GNSS is commonly used for vehicle navigation and various location-based ITS services. However, in GNSS shadow roads such as tunnels and underground highways, it is challenging to provide these services. With the rapid expansion of GNSS shadow roads, the need for radio positioning technology that can serve the role of GNSS in these areas has become increasingly important to provide accurate vehicle navigation and various location-based ITS services. This paper proposes a new GNSS shadow road radio positioning technology using multidirectional BLE beacon signals. The structure of a multidirectional BLE beacon that radiates different BLE beacon signals in two or four directions is introduced, and explains the principle of differential RSSI technology to determine the vehicle's location using these signals. Additionally, the technology used to determine the vehicle's speed is described. A testbed was constructed to verify the performance of the developed multidirectional BLE beacon-based radio navigation system. The current status and future plans of the testbed installation are introduced, and the results of position and speed experiments using the testbed for constant speed and deceleration driving are presented.

Keywords: multi-directional BLE beacon; tunnel positioning; GNSS shadow roads

1. Introduction

In outdoor environments, GNSS (Global Navigation Satellite System) is commonly used to provide various location-based ITS (Intelligent Transportation System) services such as vehicle navigation to guide driving routes to destinations, detection and warning of unexpected situations ahead, and guidance at entry and exit points, as well as precise navigation services for autonomous vehicles [1]. However, in GNSS shadow roads such as tunnels and underground highways, it is difficult to provide these services due to the inability to obtain accurate location information. Particularly in the case of GNSS shadow roads, additional services such as location-based evacuation route guidance are of significant importance due to the high fatality rate in the event of an accident on those roads.

Moreover, with the rapid expansion of tunnel road construction and the construction plans for underground highways in Korea, the need for radio positioning technology in GNSS shadow roads has become increasingly important to provide accurate vehicle navigation and various location-based ITS services.

WiFi, Cellular, BLE (Bluetooth Low Energy), and GNSS repeater technologies for radio positioning in shadow roads do not meet the accuracy requirements for vehicle navigation, and INSs (Inertial Navigation Systems) using an IMU (Inertial Measurement Unit) or odometer have the problem of cumulative errors. Radio positioning systems using GNSS pseudo-lite or special RF signals are difficult to use with smartphones or existing navigation systems, making it challenging to expand positioning services [2].

To solve these issues, a new GNSS shadow road radio positioning technology named CLOBER (Cross-shape LOcation BEacon Radio) using MD (multidirectional) BLE beacon signals has been proposed [3,4]. By radiating BLE beacon channel signals with different IDs in two or four directions, it can solve the problems of interference between beacon signals and the channel model issues in RSS (relative signal strength) of existing BLE positioning technologies. A positioning system using this BLE CLOBER has been developed and commercialized as an indoor parking lot vehicle navigation and parking route guidance system [5].

This paper proposes a technology to determine the position and speed of vehicles moving at high speed in tunnels and underground highways using DRSS (differential RSS) obtained by differencing signals radiated from MD BLE beacons. The developed BLE CLOBER-based radio positioning technology provides an accuracy of 1–3 m comparable to GNSS accuracy at both low and high speeds, and since it uses BLE signals, it can be directly applied to smartphones and vehicle navigators, making it advantageous for service expansion.

2. MD BLE Beacon-Based Radio Navigation

2.1. BLE CLOBER-Based Radio Navigation

In previous studies, positioning technologies using WiFi, Cellular, BLE, GNSS repeater, and GNSS pseudo-lite, as well as radio positioning using special RF signals such as UWB (ultra-wideband), and indoor positioning using DR (dead reckoning) have been proposed [2]. The positioning technology using WiFi and Cellular has low positioning accuracy, and the BLE positioning technology is significantly affected by interference or distance errors due to the channel environment of the propagation paths of multiple beacon signals. GNSS repeater is affected by the location of the GNSS receiver installed outside, and GNSS pseudo-lite has problems with multipath and difficulty in receiving signals on smartphones, etc. UWB positioning has high positioning accuracy, but requires separate terminals, and DR navigation technology has the problem of cumulative errors in IMU and odometer sensors, causing continuous increases in position and direction errors.

To solve these problems, the BLE CLOBER positioning technology, which radiates BLE beacon signals with different IDs in two or four directions using multidirectional antennas, has been proposed [3,4]. Figure 1 shows the signal radiation pattern of the BLE CLOBER equipped with two or four directional antennas. In the figure, each antenna radiates BLE beacon signals with different IDs and radiates signals in a time-division scheme to minimize interference. Since the signals reaching the vehicle from each antenna have very similar propagation paths, it solves the channel environment problem that causes position errors in existing BLE positioning.

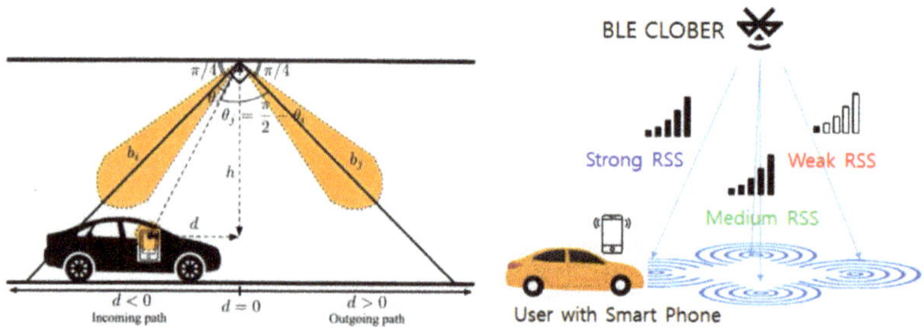

Figure 1. BLE CLOBER equipped with two or four directional antennas.

This paper proposes a technology to determine the position and speed of vehicles moving at high speeds in tunnels and underground highways using DRSS (differential RSSI) obtained by differentiating the RSS (Received Signal Strength) of signals radiated from MD BLE beacons. A vehicle passing through BLE CLOBER equipped with two directional antennas, as shown in Figure 1, has an RSS signal reception profile as shown in Figure 2, and DRSS can be obtained by differencing these signals. Since the position and antenna radiation direction of BLE CLOBER are known and stored in the navigation map, the vehicle position can be estimated using DRSS. In addition, the vehicle speed can be estimated using both the slope of the DRSS and the time taken to pass through two adjacent BLE CLOBERs.

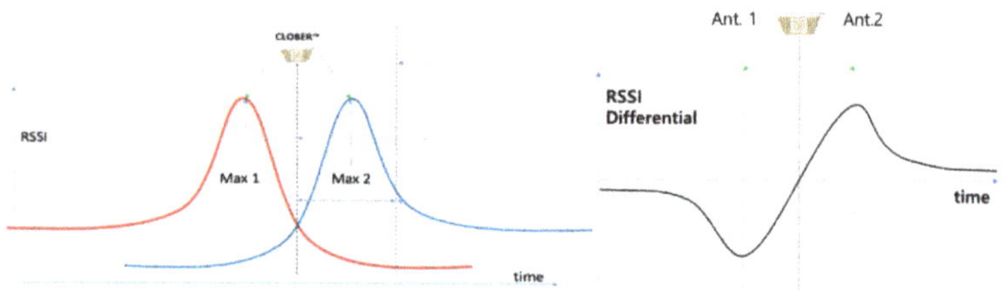

Figure 2. Received RSS profile from two-directional antennas (red from left direction and blue from right) and their DRSS profile.

2.2. BLE CLOBER Testbed Construction Status and Future Plans

To verify the high-speed performance of the BLE CLOBER positioning technology, outdoor experiments were conducted at the proving ground, and the position performance of 1–3 m, depending on the speed, was confirmed compared to the RTK position. Based on these experimental results, a testbed was constructed in the Gise tunnel located in the Daegu Technopolis area to verify the feasibility of BLE CLOBER, as shown in Figure 3. Ten BLE CLOBERs were installed at 30-m intervals, and a wired network was established for beacon maintenance, monitoring, and remote control. Using the testbed, we plan to develop position and speed estimation algorithms using BLE CLOBER in tunnel environments and analyze their accuracy. Additionally, remote control technology for beacon monitoring and maintenance will be developed. After verifying feasibility, we plan to expand the BLE CLOBER testbed by more than 1 km by the end of 2025, as shown in Figure 4, and add UWB positioning infrastructure to a 500-m section for precise performance analysis.

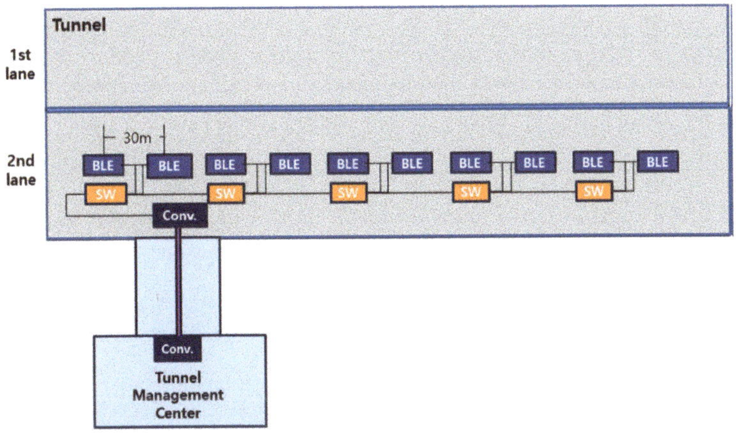

Figure 3. Architecture of the BLE CLOBER testbed in the Gise Tunnel.

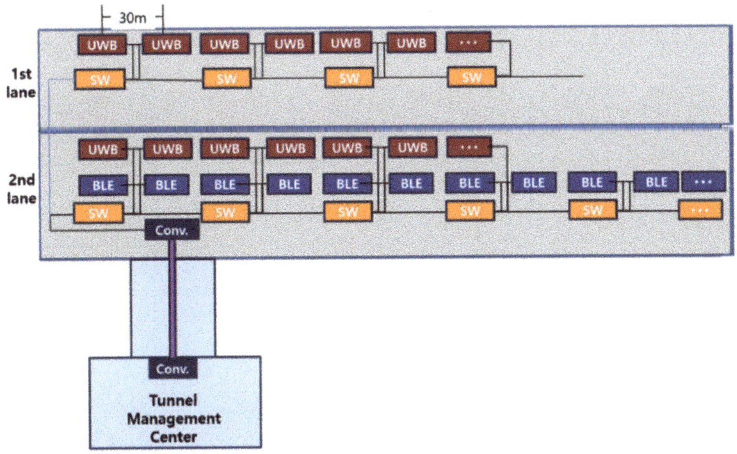

Figure 4. Future plan for the BLE CLOBER testbed in the Gise Tunnel.

3. Experiment Results

Driving experiments were conducted using the BLE CLOBER testbed installed in the Gise Tunnel. The driving experiments included constant speed tests and acceleration/deceleration tests. The constant speed tests were performed in Cruise drive mode at speeds of 40, 60, and 80 km/h to maintain a constant speed. Figure 5 shows the position and speed profiles for the constant speed experiments. In the figure for speed profile, the blue dots denote the speed computed from the slope of DRSS, and red dots are obtained from pass time between two CLOBERS. As shown in the figure, the speed estimation performance was excellent with an error within 5 km/h, and since it was a constant speed operation, the position increased linearly.

Acceleration/deceleration tests were conducted to verify the ability to detect speed changes. Figure 6 shows the position and speed profiles for the acceleration/deceleration tests. It can be observed that the speed changes were detected promptly as the vehicle passed through the CLOBER. With the future plan of expanding the testbed and the establishment of UWB precision positioning infrastructure, it is expected that more accurate positioning performance will be analyzed in the next studies.

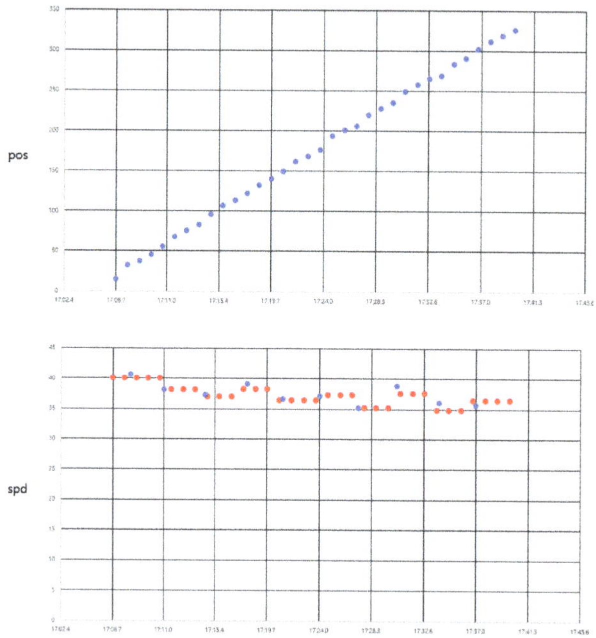

(**a**) 40 km/h speed experiment

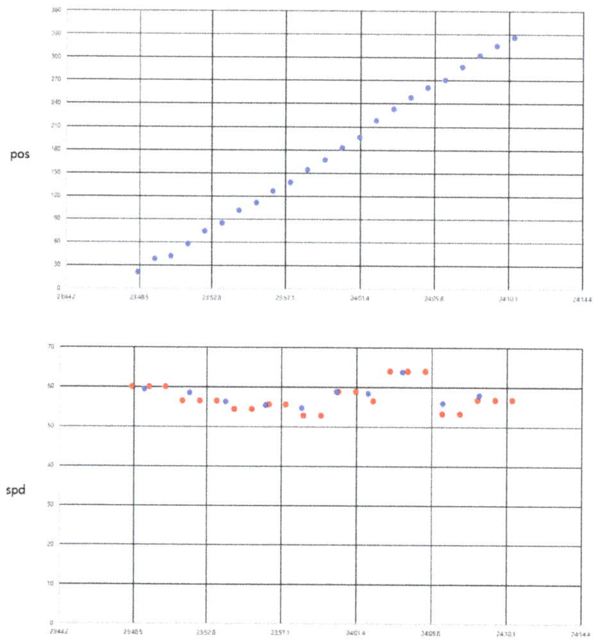

(**b**) 60 km/h speed experiment

Figure 5. *Cont.*

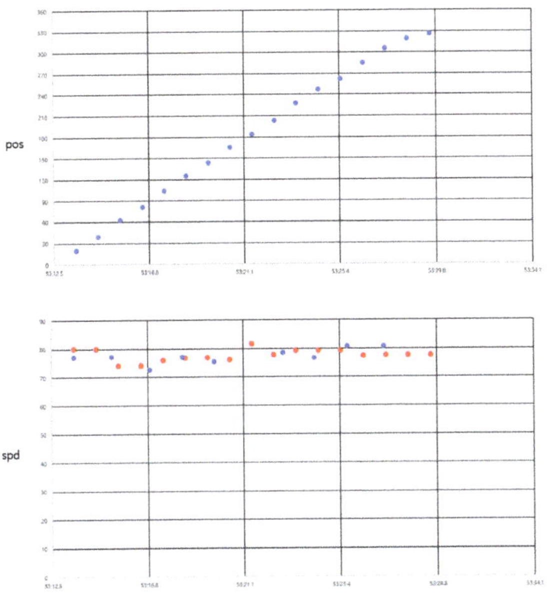

(**c**) 80 km/h speed experiment

Figure 5. Position and speed profiles for constant speed experiments: (**a**) 40 km/h speed; (**b**) 60 km/h speed; and (**c**) 80 km/h speed.

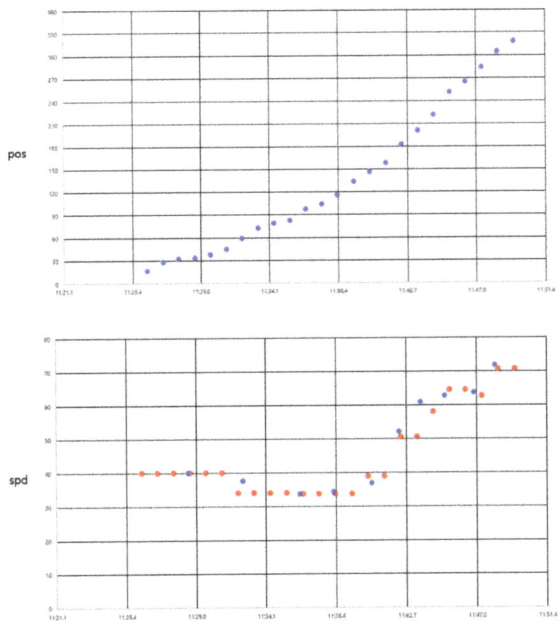

(**a**) acceleration experiment

Figure 6. *Cont.*

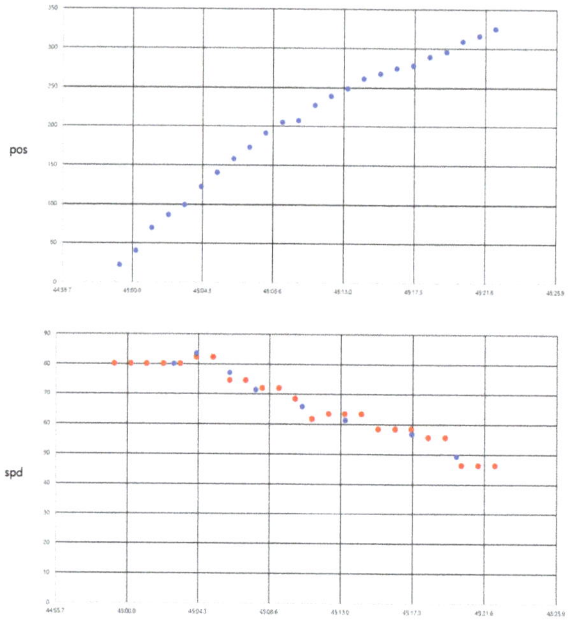

(**b**) deceleration experiment

Figure 6. Position and speed profiles for acceleration/deceleration experiments. (**a**) Acceleration. (**b**) Deceleration.

4. Conclusions

This paper proposes a technology to determine the position and speed of vehicles moving at high speeds in tunnels and underground highways using DRSS (differential RSSI) obtained by differencing signals radiated from multidirectional BLE beacons. To verify the performance of the proposed technology, constant speed experiments and acceleration/deceleration driving experiments were conducted using the BLE CLOBER testbed. The experimental results confirmed that the BLE CLOBER-based radio positioning technology developed provides an accuracy of 1–3 m at both low and high speeds, and can quickly detect speed changes. The proposed BLE CLOBER-based radio positioning technology can provide position information compatible with outdoor GNSS accuracy and can be directly applied to smartphones and vehicle navigators, allowing for rapid service expansion.

Author Contributions: Conceptualization & methodology, T.-K.S.; software, J.-Y.J.; validation & formal analysis, J.-W.K.; investigation, S.-J.K., W.-W.L.; writing—original draft preparation, T.-K.S.; writing—review and editing, T.-K.S. All authors have read and agreed to the published version of the manuscript.

Funding: This research received no external funding.

Institutional Review Board Statement: Not applicable.

Informed Consent Statement: Informed consent was obtained from all subjects involved in the study.

Data Availability Statement: Private data.

Conflicts of Interest: Author J.-W.K. and J.-Y.J. were employed by the company WiFive. The remaining authors declare that the research was conducted in the absence of any commercial or financial relationships that could be construed as a potential conflict of interest.

References

1. Alam, M.; Ferreira, J.; Fonseca, J. *Intelligent Transportation Systems*; Springer: Cham, Switzerland, 2016.
2. Bensky, A. *Wireless Positioning*; Artech House: Norwood, MA, USA, 2008.
3. Yoo, S.-H.; Sung, T.-K. Development of Precise Indoor Location System Using Multidirectional Beacon. In Proceedings of the 2017 International Conference on Indoor Positioning and Indoor Navigation (IPIN 2017), Sapporo, Japan, 18–21 September 2017.
4. Sung, T.K. Device for Estimating Location and Method for Estimating Location by Using Downlink Accesspoint. U.S. Patent 9,942,709, 10 April 2018.
5. Park, P.; Marco, P.D.; Santucci, F.; Jung, M.; Sung, T.-K. Multidirectional Differential RSS Technique for Indoor Vehicle Navigation. *IEEE Internet Things J.* **2023**, *10*, 241–253. [CrossRef]

engineering proceedings

MDPI

Proceeding Paper

A Study on High-Precision Vehicle Navigation for Autonomous Driving on an Ultra-Long Underground Expressway †

Kyoung-Soo Choi [1,*], Yui-Hwan Sa [1], Min-Gyeong Choi [1], Sung-Jin Kim [1] and Won-Woo Lee [2]

[1] Autonomous Driving Technology Research Division, ITS R&D Department, Korea Automotive Technology Institute, Cheonan-si 31214, Republic of Korea; yhsa@katech.re.kr (Y.-H.S.); mgchoi@katech.re.kr (M.-G.C.); sjkim@katech.re.kr (S.-J.K.)

[2] Digital Convergence Research Division, Korea Expressway Corporation Research Institute, Hwaseong-si 18487, Republic of Korea; wonwoo2.lee@gmail.com

* Correspondence: kschoi@katech.re.kr

† Presented at the 2025 Suwon ITS Asia Pacific Forum, Suwon, Republic of Korea, 28–30 May 2025.

Abstract

GPSs typically have an accuracy ranging from a few meters to several tens of meters. However, when corrected using various methods, they can achieve an accuracy of several tens of centimeters. In autonomous driving, a positioning accuracy of less than 50 cm is required for lane-level positioning, route generation, and navigation. However, in environments where GPS signals are blocked, such as tunnels and underground roads, absolute positioning is impossible. Instead, relative positioning methods integrating IMU, IVN, and cameras are used. These methods are influenced by numerous variables, however, such as vehicle speed and road conditions, resulting in lower accuracy. In this study, we conducted experiments on current vehicle navigation technologies using an autonomous driving simulation vehicle in the Suri–Suam Tunnel of the Seoul Metropolitan Area 1st Ring Expressway. To recognize objects (lane markings/2D/3D) for position correction inside the tunnel, data on tunnel and underground road infrastructure in Seoul and Gyeonggi Province was collected, processed, refined, and trained. Additionally, a Loosely Coupled-based Kalman Filter was designed and applied for the fusion of GPSs, IMUs, and IVNs. As a result, an error of 113.62 cm was observed in certain sections. This suggests that while the technology is applicable for general vehicle lane-level navigation in ultra-long tunnels spanning several kilometers for public service, it falls short of meeting the precision required for autonomous driving systems, which demand lane-level accuracy. Therefore, it was concluded that infrastructure-based absolute positioning technology is necessary to enable precise navigation inside tunnels.

Keywords: K-underground expressway; autonomous driving; vehicle navigation; U-GPS; BLE; UWB

1. Introduction

Performance experiments of navigation technology based on an autonomous driving simulation vehicle were conducted in the Suri Tunnel (approximately 1.8 km) and the Suam Tunnel (approximately 1.2 km) of the Seoul Metropolitan Area 1st Ring Expressway. The test vehicle used was a EV6(KIA, Seoul, Republic of Korea) configured to enable a hybrid navigation system that integrates GPS-based absolute positioning with relative positioning using IMU (Inertial Measurement Unit), IVN (In-Vehicle Network), a camera, and LiDAR. The detailed specifications of the system are shown in Figure 1 below.

Sensor	Specification	Quantity
GPS (Trimble)	RTK : 0.015m + 1ppm 50Hz Pos. Output	1 EA
IMU (OXTS)	Roll/Pitch : 0.03° 1 σ Heading : 0.1° 1 σ 100Hz Output	1 EA
Camera (FLIR)	Resolution : 4096 x 3000	4 EA
LiDAR (OUSTER)	CH : 128 Range : 200m VFOV : 45°	1 EA
Object Detection (Static / Dynamic)	81.8 mAP / 86.32 mAP	1 Set
DAQ	Intel Core I7-5700 16GB RAM ..	1 EA

Figure 1. System Configuration and Specification.

To enable object recognition inside tunnels, data collection, refinement, and preprocessing were conducted for tunnels and underground roads in the Seoul and Gyeonggi regions, covering a total length of approximately 30 km. For relative position correction, a total of 18 types of tunnel interior objects were selected, including firefighting equipment (3 types), alarm systems (11 types), and evacuation facilities (4 types).

For the implementation of Lane/Object (2D/3D) Detection [1–3], the following models were used: YOLOP for Lane Detection, YOLOv9 for 2D Object Detection [2], and CasA for 3D Object Detection [1]. The objects and models used for training, as well as their results, are shown in Figure 2 below.

Class	Object Name	Installation Interval
Firefighting Equipment	Fire Extinguishing Device	Within 50m
	Indoor Fire Hydrant System	Within 50m
	Water Spray System	25–50m
Alarm Systems	Emergency Alarm System	Within 50m
	Automatic Fire Detection System	Within 50m
	Emergency Broadcasting System	Within 50m
	Emergency Telephone	Within 250m
	CCTV	Inside Tunnel: 200–400m Outside Tunnel: Within 500m
	Automatic Incident Detection System	Video-based Detection: Around 100m Unexpected Situation Monitoring: Reflecting Tunnel Characteristics
	Rebroadcasting System	Installed to enable listening within tunnel sections
	Tunnel Entrance Information Display	Within 500m before the tunnel
	Entry Restriction Signboard	Every 50m up to 200m before the tunnel entrance
	Tunnel Entry Restriction System	-
	Lane Usage Control Signal	400 – 500m
Evacuation Facilities	Emergency Lighting	Installed in conjunction with nighttime lighting circuits
	Guide Lights (A, B)	Distance Indicator: Approximately every 50m Evacuation Exit Indicator: Every 250m
	Refuge Facilities	Within 250–300m

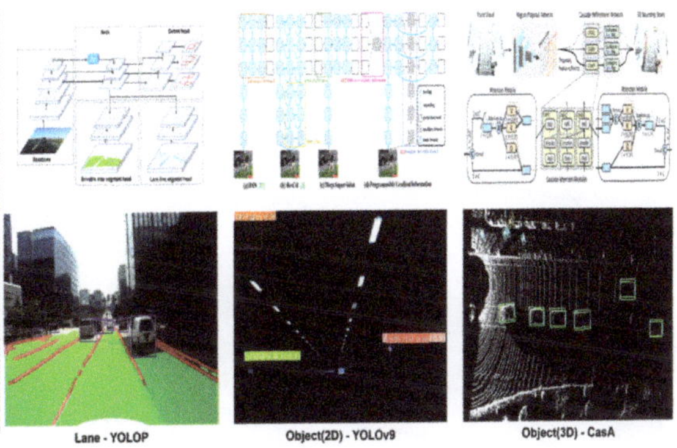

Figure 2. Objects and Recognition Models/Results for Relative Position Correction in Tunnels.

To integrate GPS, IMU, and IVN, a Loosely Coupled-based Kalman Filter was designed and applied [4,5]. An error model was formulated and incorporated, considering factors such as DR (Dead Reckoning) attitude angles, speedometer conversion coefficients, and gyroscope bias. The detailed implementation is shown in Figure 3 below.

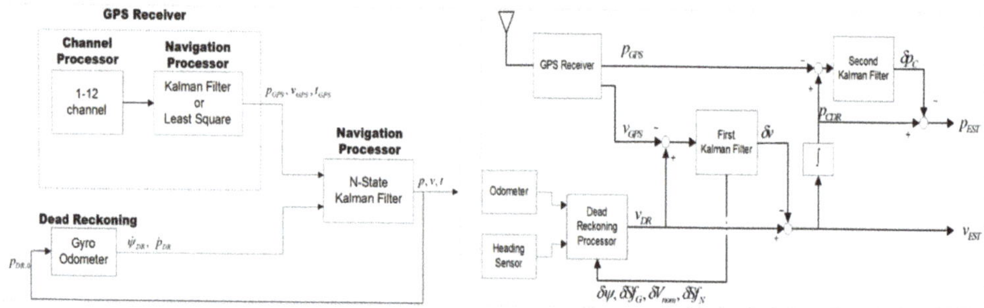

Figure 3. GPS/IMU/IVN Fusion Filter.

Based on the configured system, an experiment was conducted by driving round-trip from Anyang–Pyeongchon to Pangyo, passing through the Suam Tunnel and Suri Tunnel in sequence (Figure 4). The driving conditions followed the expressway speed regulations and were conducted on the fourth lane after 4:00 p.m. The driving results are presented in Figure 5 below.

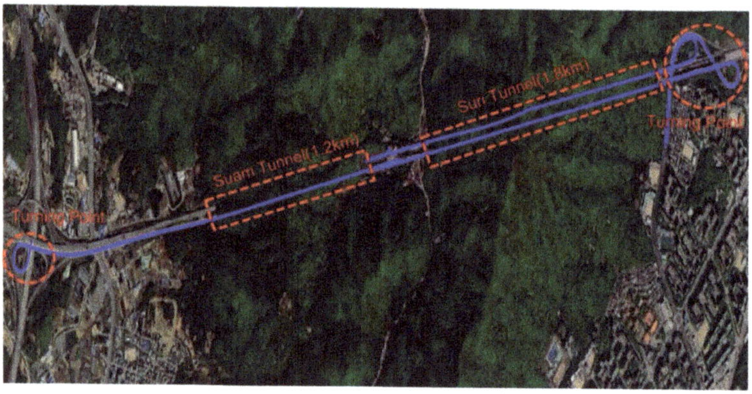

Figure 4. Driving Route.

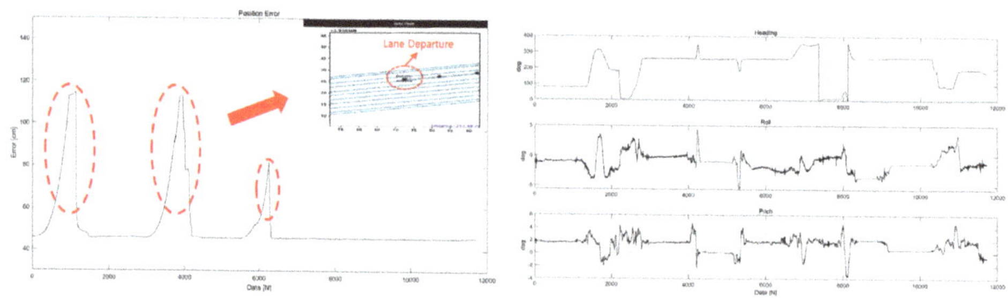

Figure 5. Driving Test Results.

2. Results

In this study, a positioning accuracy experiment was conducted on vehicle navigation technology in ultra-long underground expressways using an autonomous driving simulation vehicle. In occlusion sections where GPS signals are interrupted, inertial navigation positioning was performed by fusing IMU/IVN, and vision-based navigation positioning was conducted by integrating a camera/LiDAR. Due to challenging tunnel environments (e.g., signal obstruction by moving objects and multipath fading), fluctuations in absolute positioning were observed in certain sections, resulting in a maximum error of 113.62 cm. While this level of accuracy is acceptable for non-research vehicles using commercial navigation systems, it was deemed insufficient for application in autonomous driving systems. Based on the experimental results, it was concluded that to achieve accurate positioning even in GPS-denied areas for both autonomous vehicles and non-research vehicles, a continuous absolute positioning system integrating infrastructure-based technologies such as virtual GPS signals, BLE (Bluetooth Low Energy), and UWB (Ultra-Wideband) is required. Future research will focus on integrating these technologies with autonomous driving systems and commercial navigation systems to develop a stable vehicle navigation system in way tunnel environments.

Author Contributions: Conceptualization, K.-S.C. and S.-J.K.; methodology, K.-S.C.; software, Y.-H.S. and M.-G.C.; validation, K.-S.C. and Y.-H.S.; formal analysis, M.-G.C.; investigation, K.-S.C. and M.-G.C.; resources, S.-J.K. and W.-W.L.; data curation, K.-S.C. and Y.-H.S. and M.-G.C.; writing—original draft preparation, K.-S.C. and Y.-H.S.; writing—review and editing, K.-S.C. and Y.-H.S.; visualization, M.-G.C.; supervision, S.-J.K. and W.-W.L.; project administration, S.-J.K. and W.-W.L.; funding acquisition, S.-J.K. and W.-W.L. All authors have read and agreed to the published version of the manuscript.

Funding: This research was funded by the Ministry of Land, Infrastructure and Transport / Korea Agency for Infrastructure Technology Advancement under the project titled "Development of Technology to Enhance Safety and Efficiency of Ultra-Long K-Underground Expressway Infrastructure" [Project No. RS-2024-00416524]. The APC was funded by the same project.

Institutional Review Board Statement: Not applicable.

Informed Consent Statement: Not applicable.

Data Availability Statement: Data are available in this manuscript.

Conflicts of Interest: The authors declare no conflict of interest.

References

1. Wu, H.; Deng, J.; Wen, C.; Li, X.; Wang, C.; Li, J. CasA: A Cascade Attention Network for 3-D Object Detection from LiDAR Point Clouds. IEEE Trans. Geosci. *Remote Sens.* **2022**, *60*, 5704511. [CrossRef]
2. Wang, C.Y.; Yeh, I.H.; Liao, M. Learning What You Want to Learn Using Programmable Gradient Information. In Proceedings of the Computer Vision—ECCV, Milan, Italy, 29 September–4 October 2024.
3. Wu, D.; Liao, M.; Zhang, W.; Wang, X.; Bai, X.; Cheng, W.; Liu, W.Y. A simple feature augmentation for domain generalization. In Proceedings of the IEEE/CVF International Conference on Computer Vision Workshops, Montreal, QC, Canada, 1 October 2021.
4. Groves, P.D. *Principles of GNSS, Inertial, and Multi-Sensor Integrated Navigation Systems*, 2nd ed.; Artech House: Norwood, MA, USA, 2008.
5. Titterton, D.H.; . Weston, J.L. *Strapdown Inertial Navigation Technology*, 2nd ed.; The Institution of Engineering and Technology: London, UK, 2004.

engineering proceedings

MDPI

Proceeding Paper

Comparing Dynamic Traffic Flow Between Human-Driven and Autonomous Vehicles Under Cautious and Aggressive Vehicle Behavior †

Maftuh Ahnan and Dukgeun Yun *

Department of Highway & Transportation Research, Korea Institute Civil Engineering & Building Technology (KICT) School, University of Science & Technology (UST), 283 Goyang-daero, Ilsanseo-gu, Goyang-si 10223, Gyeonggi-do, Republic of Korea; maftuh@kict.re.kr

* Correspondence: dkyun@kict.re.kr

† Presented at the 2025 Suwon ITS Asia Pacific Forum, Suwon, Republic of Korea, 28–30 May 2025.

Abstract

This study explores the impact of driving behaviors, specifically cautious and aggressive, on the performance of human-driven vehicles (HDVs) and autonomous vehicles (AVs) in traffic flow dynamics. It focuses on various metrics, including level of service (LOS), average speed, traffic volume, queue delays, carbon emissions, and fuel consumption, to assess their effects on overall performance. The findings reveal significant differences between cautious and aggressive AVs, particularly at varying market penetration rates (MPRs). Aggressive autonomous vehicles demonstrate greater traffic efficiency compared to their cautious counterparts. They achieve higher levels of service, improving from poor performance at low MPRs to significantly better performance at higher MPRs and in fully autonomous scenarios. In contrast, cautious AVs often experience poor service ratings at low MPRs, with an improvement in performance only at higher MPRs. Regarding environmental performance, aggressive AVs outperform cautious ones in terms of reduced emissions and fuel consumption. The emissions produced by aggressive AVs are significantly lower than those from cautious AVs, and they further decrease as the MPRs increases. Additionally, aggressive AVs show a considerable reduction in fuel usage compared to cautious AVs. While cautious AVs improve slightly at higher MPRs, they continue to generate higher emissions and consume more fuel than their aggressive counterparts. In conclusion, aggressive AVs offer better traffic efficiency and environmental performance than both cautious AVs. Their ability to improve road efficiency and reduce congestion positions them as a valuable asset for sustainable transportation. Strategically incorporating aggressive AVs into transportation systems could lead to significant advancements in traffic management and environmental sustainability.

Keywords: autonomous vehicles; human driven vehicles; aggressive; cautious

1. Introduction

Human-driven cars add unpredictability, which may affect autonomous vehicle traffic networks. For future urban planning and infrastructure development, understanding how autonomous cars affect traffic dynamics, especially human-operated vehicles, is crucial [1–4] Research has shown several outcomes throughout this transitional phase [5–7]. As autonomous cars become more common, their predictive driving and sensor-driven navigation systems will improve traffic flow [8,9]. Simulations show that autonomous cars might minimize traffic congestion and conflicts by improving lane-keeping and deceleration management [10,11]. However, low penetration rates cause many issues because traditional vehicles lack the predictability of autonomous vehicles, undermining the benefits of automated vehicles in a fully automated setting [12–14].

In 2017, the European Union launched the CoEXist effort to prepare cities for the coexistence of autonomous cars (AVs) and human-driven vehicles (HDVs) on the roads [15]. The effort aims to gather data on driver behavior and the impact of autonomous vehicles on mixed traffic scenarios. Based on the insights from the CoEXist research, VISSIM version 2022 provided three driving logics for autonomous cars to replicate their behavior in a simulated environment. There exist three categories of driving behaviors: cautious, normal, and aggressive. Each one delineates a distinct sort of conduct in both the longitudinal and lateral dimensions.

This research simulates many scenarios to examine the market penetration of autonomous cars, highlighting traffic flow and demand in relation to human-operated vehicles. This study used VISSIM version 2022, an intricate microscopic traffic simulation software, to examine the impact of aggressive and cautious behaviors of self-driving cars on traffic flow on Jakarta Harbour Tollroad. This study evaluates traffic volume, efficiency, and accident reduction in mixed traffic conditions. The model of autonomous vehicles is based on the CoExist Project, which is given by VISSIM version 2022. On the other hand, the model of human-driven vehicles is based on the actual traffic circumstances that occur on the Jakarta Harbor Toll road.

2. Literature Review

Multiple studies have forecasted the market penetration rates (MPR) of autonomous vehicles (AVs). By 2045, studies project that the MPR of driverless cars would range from 24 to 87 percent. Despite the correlation between the intrinsic advantages of autonomous cars and complete market saturation, achieving a market penetration of 100 percent remains a significant challenge [15–17]. Throughout the transitional phase, autonomous vehicles (AVs) will engage with human-driven vehicles (HDVs). Researchers anticipate that the intricate environment generated by these vehicles, with differing levels of autonomy, would affect driving behaviors within traffic dynamics. This study examines traffic dynamics, human driver behavior, and the integration of autonomous cars based on the CoExist Project, utilizing cautious and aggressive behaviors to understand the challenges and consequences of integrating autonomous and human-driven vehicles. We will examine human driving behaviors to guarantee safe and efficient vehicle operation on highways. We adjust the forceful driving behaviors of both autonomous and human-operated cars to correspond with their surroundings. Thereafter, we utilize VISSIM version 2022 to assess the mobility, safety, and adoption rate of these vehicles.

3. Methodology

3.1. Research Area

This study examines the Harbour Tollroad in Jakarta, Indonesia. Pedicabs and motorcyclists are barred from accessing the toll road, with only autonomous vehicles, cars, Heavy Goods Vehicle (HGV) and buses authorised for entry. A single on-ramp and a single off-ramp connect each road. Figure 1 visually illustrates the location of the research study.

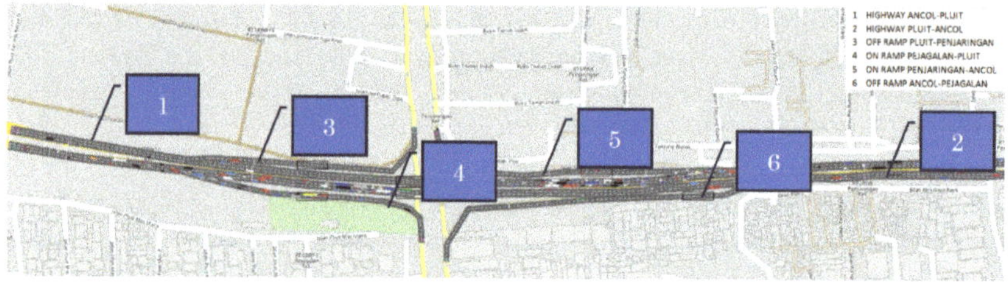

Figure 1. The research area relevant to this study.

3.2. Research Design Model Development

This research allows modifications based on the traffic conditions we assess. This traffic composition study primarily examines autonomous vehicles (AVs) and human-operated vehicles (HOVs). Utilizing parameter values from the CoEXist study, the AV aggressive and cautious behavior classifications align with the fundamental principles of VISSIM version 2022. One may choose from several situations and market penetration rates (MPRs). The scholarly literature utilizes a similar concept by using maximum road capacity to accurately represent probable peak traffic scenarios. The scenario approach juxtaposes a dynamic examination of human-operated cars with analyses of both aggressive and cautious autonomous vehicles. Each scenario will conduct three essential simulations, specifically as follows:

1. A baseline is established by this scenario, which simulates typical traffic in the absence of autonomous vehicles. This scenario is referred to as human-driven vehicle traffic.
2. Market penetration rates (MPRs) for autonomous vehicles can range anywhere from 10% to 90% in increments of 10%, depending on the level of competition in the environment. Mixed traffic environments are characterized by this.
3. Fully Autonomous Traffic: Autonomous vehicles constitute one hundred percent of the traffic, exemplifying a potential future scenario in which they predominate road usage.

The simulations generate various outputs regarding performance and traffic flow efficiency. We use these outputs to analyze the procedure's results. Figure 2 illustrates the three fundamental stages previously mentioned, together with the processes associated with each level.

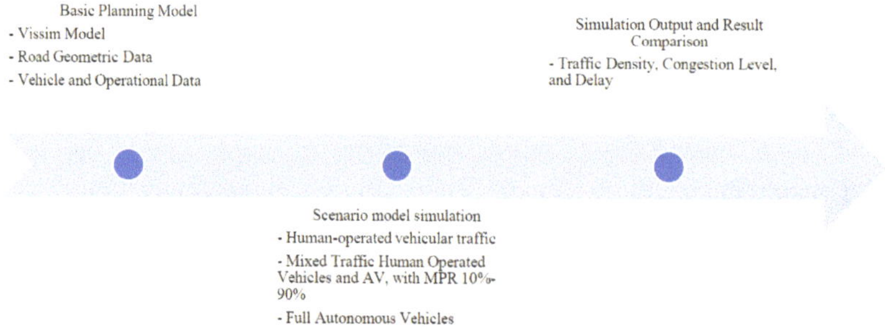

Figure 2. Three phases of analyzing dynamic mixed traffic comprising human-driven vehicles with aggressive and cautious autonomous vehicles.

We identified the key input factors that significantly affect vehicle performance, including their minimum and maximum values. Table 1 presents the findings. We established the range values of the input variables using the minimum and maximum values derived from prior studies.

Table 1. Parameters for human driven vehicle behavior.

Parameters	Park et al. [18]	He [19]	Shin et al. [12]	Current Research
Following model				
Look-ahead distance	200~300	72~108	-	60~100
Look-back distance	200~300	72~108	-	60~100
Interaction of objects	1~5	-	-	2
Interaction of vehicles	-	-	-	1
Car following model				
CCO, Distance to standstill (m)	1.0~2.0	0.0~20.0	0.0~3.0	5.00
CC1, Time for headway (s)	0.5~3.0	0.0~5.0	0.0~5.0	5.00
CC2, Distance maintained between vehicles (m)	0.0~15.0	0.0~10.0	0.0~40.0	40.0
CC3, Criteria for entry into the subsequent (s)	−30.0~0.0	−20.0~0.0	-	−30.0
CC4, Negative subsequent threshold (m/s)	−1.0~0.0	−5.0~0.0	-	−1.0
CC5, Positive subsequent threshold (m/s)	0.0~1.0	0.1~5.0	-	5.0
CC6, Distance-dependent oscillation (1/(m/s))	0.0~20.0	0.1~20.0	-	20.0
CC7, Acceleration oscillation (m/s^2)	0.0~1.0	−1.0~1.0	-	1.0
CC8, Static acceleration (m/s^2)	1.0~8.0	0.0~8.0	-	8.0
CC9, 80 km/h acceleration (m/s^2)	0.5~3.0	0.0~8.0	-	8.0
Model of lane change				
Maximum deceleration	−5.0~−1.0	−5.0~0.0	−5.0	−5.0~−1.0
−1 m/s^2 per distance	100	100	100	100
Deceleration accepted (m/s^2)	−3.0~0.2		-	−3.0
Waiting period prior to diffusion	-	-	-	60.00
Minimum front/rear clearance (m)	0.1~0.9		-	0.50
Factor of safety distance decrease	-		0.0~1.0	0.0~1.0
Cooperative braking maximum deceleration (m/s^2)	−5.0~−1.0		−5.0~0.0	−5.0

4. Result and Discussion

4.1. Comparison of Dynamic Traffic Flow Between Human-Driven Vehicles and Autonomous Vehicles

A comprehensive comparison assesses the dynamic traffic flow characteristics of human-driven vehicles (FHDVs) and autonomous vehicles (AVs) under both cautious and aggressive behaviors. The metrics evaluated encompass level of service (LOS), mean speed, mean queue delay, mean queue length, mean emissions, and mean fuel consumption.

4.2. Level of Service (LOS)

Level of service (LOS) serves as a critical metric for assessing traffic flow efficiency, quantifying the extent of congestion on roadways. Aggressive autonomous vehicles (AVs) markedly improve the level of service (LOS) performance in comparison to cautious AVs and human-driven vehicles (FHDVs). At lower market penetration rates (MPRs), aggressive autonomous vehicles exhibit level of service E or 5, which indicates moderate to heavy congestion. As the MPRs increases, aggressive AVs show significant enhancement, achieving LOS C or 3 at an MPR of 90%, indicating decreased congestion and improved traffic flow.

Cautious AVs exhibit suboptimal performance at low MPRs, frequently resulting in LOS F or 6, indicative of significant congestion. The suboptimal performance is linked to conservative driving behaviors, including larger vehicle gaps and a cautious approach to acceleration. At elevated MPR, cautious AVs demonstrate improvement, achieving LOS C or 3 at an MPR of 90%; however, their performance remains markedly inferior to that of aggressive AVs. The findings highlight the advantages of aggressive autonomous vehicles in enhancing road capacity and minimizing delays, positioning them as essential elements of effective future traffic systems. Figures 3 and 4 illustrate the comparative results of LOS for cautious and aggressive AV behaviors.

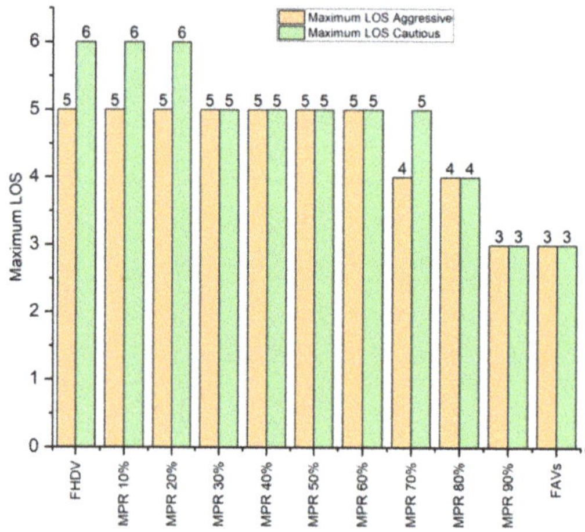

Figure 3. Graph comparing level of service cautious and aggressive autonomous vehicles.

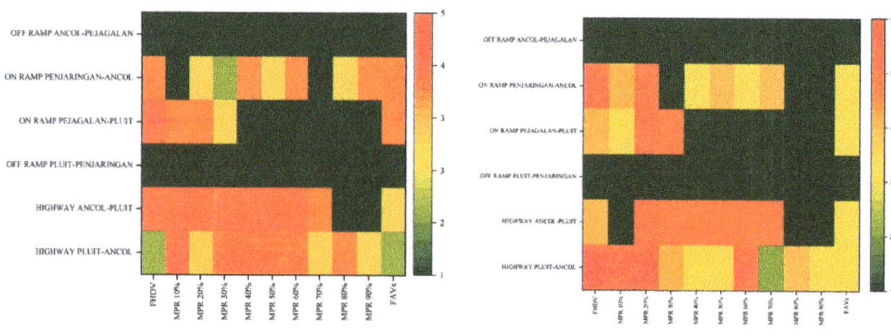

Figure 4. Heatmaps comparing level of service for aggressive (**left**) and cautious (**right**) autonomous vehicles.

4.3. Average Speed

Average speed serves as a critical measure of the efficiency of traffic flow. Aggressive autonomous vehicles consistently demonstrate superior average speeds compared to both cautious autonomous vehicles. At an MPR of 10%, the average speed of aggressive AVs is 37 km/h, which increases consistently with rising MPRs. In scenarios involving fully autonomous vehicles (FAVs), aggressive autonomous vehicles (AVs) attain a maximum average speed of 67 km/h indicating a significant enhancement as the share of autonomous vehicles in traffic rises.

Cautious autonomous vehicles exhibit reduced speeds in comparison to aggressive autonomous vehicles. At an MPR of 20%, the average speed decreases to 32 km/h, which is slower than that of FHDVs. With an increase in the MPR, cautious autonomous vehicles demonstrate a degree of improvement, achieving speeds of 63 km/h in fully autonomous conditions. The cautious characteristics of these AVs, including slower acceleration, increased headways, and a hesitance to change lanes aggressively, constrain their speed. The reduced performance significantly impacts their overall traffic efficiency. The results show that aggressive autonomous vehicles, by staying at high speeds, improve traffic flow and shorten travel times, while cautious autonomous vehicles make traffic less efficient. Figures 5 and 6 illustrate the results of comparing average speeds between cautious and aggressive AV behaviors.

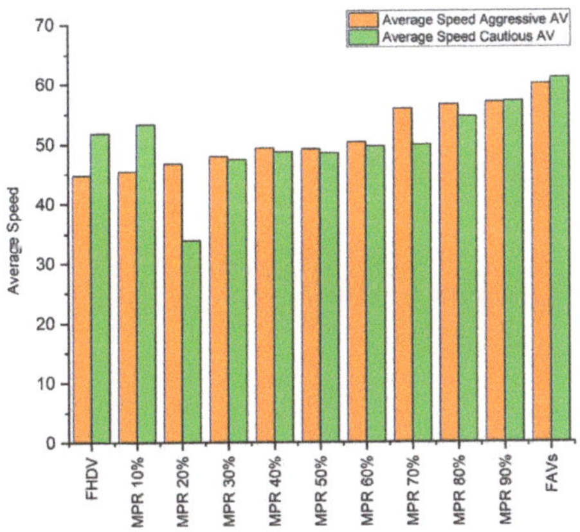

Figure 5. Graph showing comparison of average speeds between cautious and aggressive AV.

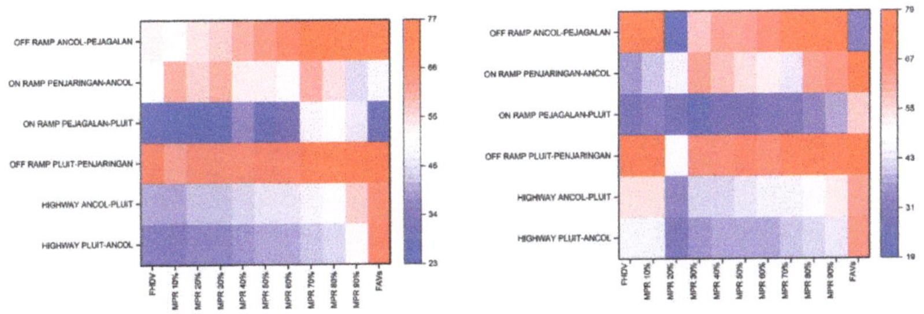

Figure 6. Heatmap comparing average speed of aggressive (**left**) and cautious (**right**) autonomous vehicles.

4.4. Maximum Volume

Maximum volume is a crucial performance metric indicating the road's capacity to accommodate traffic. Aggressive autonomous vehicles demonstrate superior performance in managing elevated traffic volumes relative to both cautious autonomous vehicles and fully human-driven vehicles. The maximum volume for FHDVs is 1231 vehicles per hour. Aggressive AVs exhibit a significant increase in maximum volume, achieving 6187 vehicles per hour in fully autonomous scenarios. While this is happening, cautious autonomous vehicles are experiencing an increase in MPR, but their performance is still lower than that of aggressive autonomous vehicles. At a market penetration rate of 20%, the maximum volume is 1586 vehicles per hour, which increases to 4329 vehicles per hour in fully autonomous scenarios. Cautious autonomous vehicles have increased headway and decreased movement due to cautious driving behaviours. The comparison of aggressive and cautious autonomous vehicles highlights the significance of driving behavior in optimizing traffic volume and road capacity utilization. Aggressive AVs demonstrate enhanced effectiveness in optimizing road usage, mitigating congestion, and improving overall traffic efficiency. Figures 7 and 8 illustrate the comparative results of maximum- capacity traffic volumes for cautious and aggressive AV behaviors.

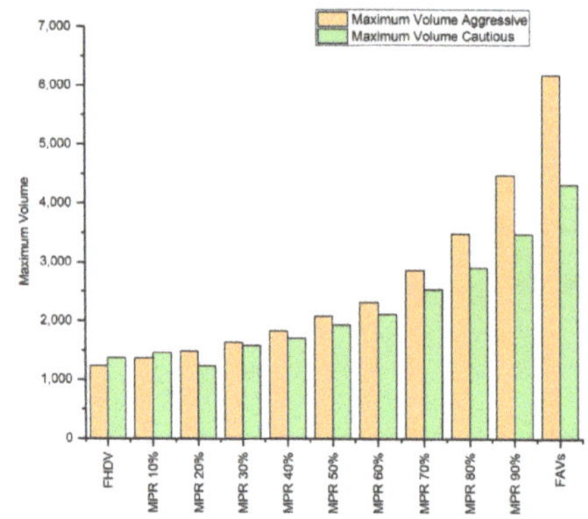

Figure 7. Graph comparing maximum-capacity traffic volumes for cautious and aggressive behaviors.

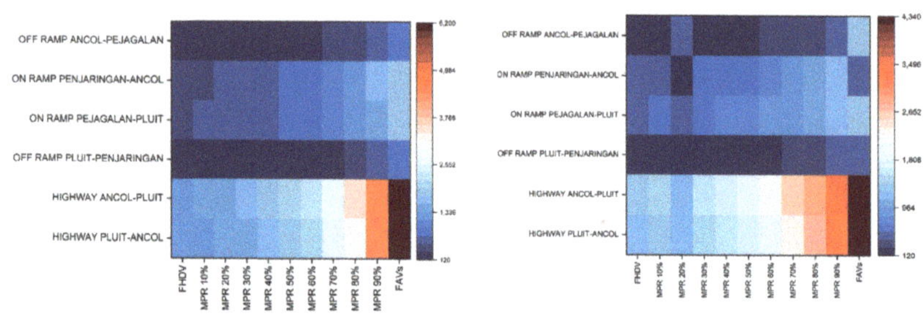

Figure 8. Heatmaps comparing traffic volume of aggressive (**left**) and cautious (**right**) autonomous vehicles.

4.5. Queue Delay and Queue Length

Queue delay and queue length serve as critical metrics for assessing congestion. Aggressive AVs markedly decrease both metrics. At FHDV levels, the mean queue delay is 40 s, reducing to 14 s at an MPR of 90%. The queue length decreases from 26 vehicles to 19 vehicles. The enhancements are linked to the capacity of aggressive autonomous vehicles to navigate intersections effectively and mitigate bottlenecks, facilitated by their dynamic driving behaviors and quicker response times.

Cautious AVs demonstrate increased delays and extended queues, especially at lower MPR. At a market penetration rate of 20%, the average queue delay reaches a maximum of 78 s, while queue lengths attain 60 vehicles. Cautious autonomous vehicles demonstrate enhanced performance with rising MPR; however, they remain inferior to aggressive autonomous vehicles. At a maximum permissible rate of 90%, the average delay decreases to 24 s, and the queue length diminishes to 38 vehicles. The cautious behavior of autonomous vehicles, marked by yielding, increased spacing between vehicles, and conservative acceleration, results in elevated delays. This underscores the necessity of implementing more assertive AV strategies to mitigate congestion and enhance traffic flow. Figures 9–11 illustrate the comparative results of queue length and delay for cautious and aggressive AV behaviors.

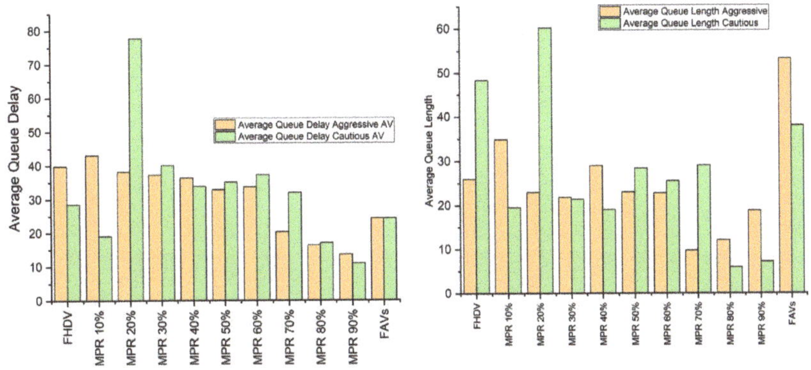

Figure 9. Graphs comparing queue delay and queue length for cautious and aggressive autonomous vehicles.

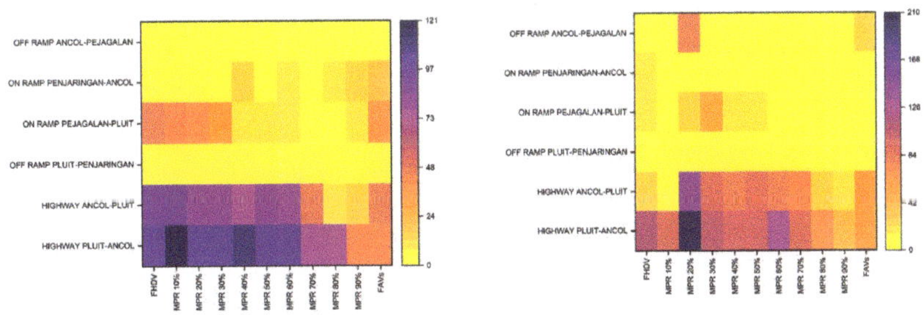

Figure 10. Heatmaps comparing queue delay for aggressive (**left**) and cautious (**right**) autonomous vehicles.

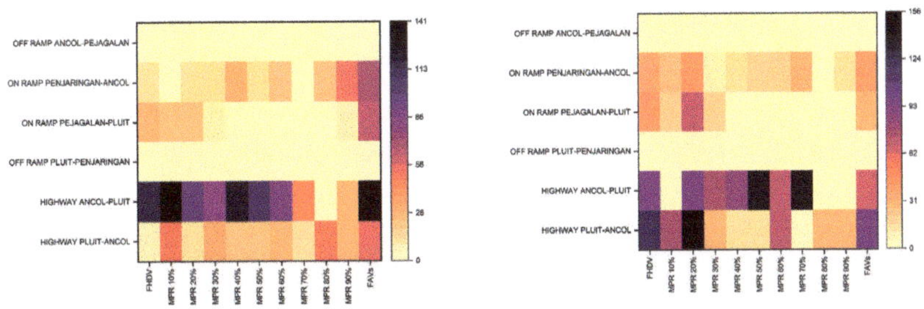

Figure 11. Heatmaps comparing queue length for aggressive (**left**) and cautious (**right**) autonomous vehicles.

4.6. Carbon Emissions

Emissions serve as a vital environmental indicator that signifies the sustainability of transportation systems. Aggressive autonomous vehicles demonstrate substantial enhancements in emission reduction. Emissions for FHDVs begin at 4286 g CO_2 and decrease to 1800 g CO_2 at an MPR of 90%. The reduction is mainly attributed to improved traffic flow, decreased idling times, and increased average speeds, which are indicative of aggressive driving behaviors. The dynamic approach of aggressive autonomous vehicles contributes to reduced fuel consumption and enhanced environmental efficiency. Cautious autonomous vehicles exhibit increased emissions at reduced market penetration rates. At a marginal abatement cost of 20%, emissions reach a maximum of 6809 g of CO_2. As the MPR increases, emissions decline, ultimately reaching 2613 g CO_2 at an MPR of 90%. As the proportion of autonomous vehicles increases, emissions tend to improve; however, cautious autonomous vehicles still result in higher emissions than their aggressive counterparts. The comparison indicates that aggressive AVs exhibit a distinct advantage regarding environmental sustainability, leading to diminished carbon footprints and enhanced fuel efficiency. Figures 12 and 13 illustrate the comparative results of carbon emissions for cautious and aggressive AV behaviors.

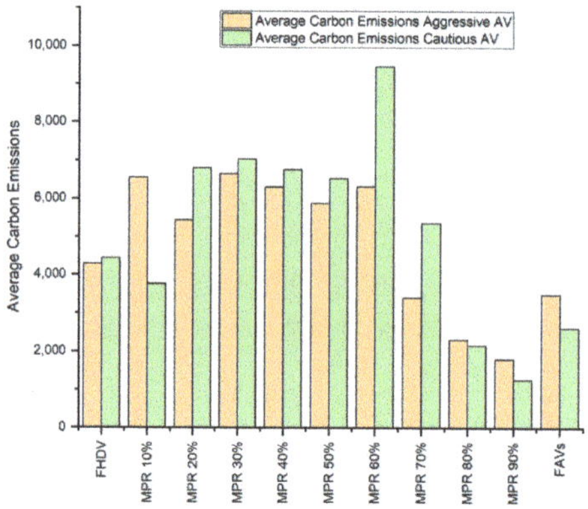

Figure 12. Graph comparing carbon emissions of cautious and aggressive autonomous vehicles.

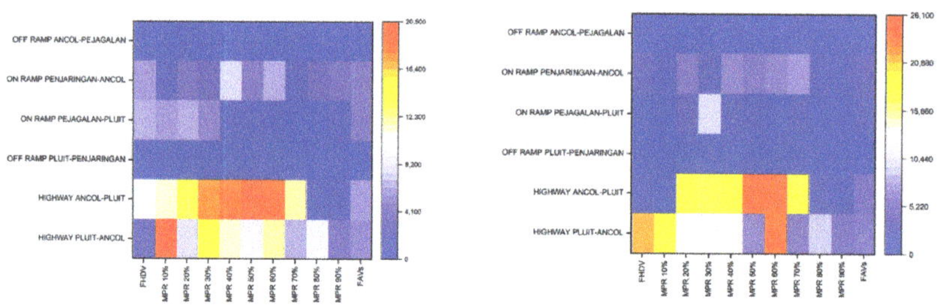

Figure 13. Heatmaps comparing carbon emissions of aggressive (**left**) and cautious (**right**) autonomous vehicles.

4.7. Fuel Consumption

Fuel consumption, strongly linked to emissions, demonstrates a notable reduction in aggressive autonomous vehicles. As depicted in Figures 14 and 15, under the full human-driven vehicles (FHDVs) condition, the fuel consumption of cautious autonomous vehicles is approximately 64 liquid gallons, which is only slightly higher than that of aggressive autonomous vehicles at 61 liquid gallons. However, upon achieving a market penetration rate of fully autonomous vehicles, aggressive autonomous vehicles exhibit a lower consumption rate of 38 liquid gallons compared to cautious autonomous vehicles. This reduction is attributable to a more gradual acceleration pattern, lower idling duration, and enhanced traffic flow efficiency. The assertive driving style of aggressive autonomous vehicles enhances fuel economy, thereby minimizing fuel waste. The elevated fuel consumption of cautious autonomous vehicles is due to frequent stops, reduced acceleration rates, and suboptimal lane management. The comparison highlights that aggressive AVs exhibit greater fuel efficiency, especially at elevated MPR, thereby serving as a crucial element of sustainable transportation systems. Figures 14 and 15 illustrate the comparative fuel consumption results for cautious and aggressive AV behaviors.

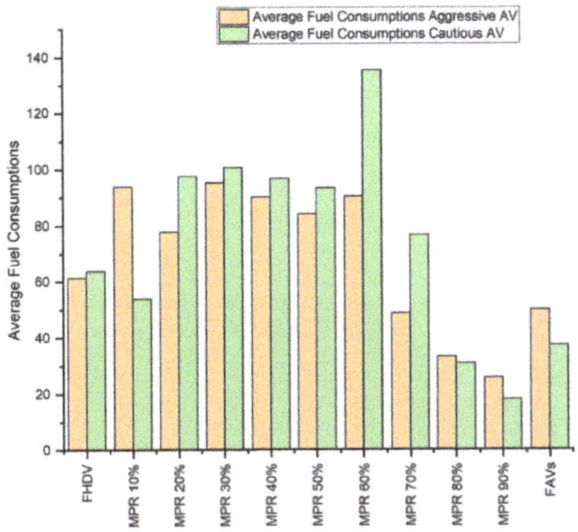

Figure 14. Graph comparing fuel consumption of cautious and aggressive autonomous vehicles.

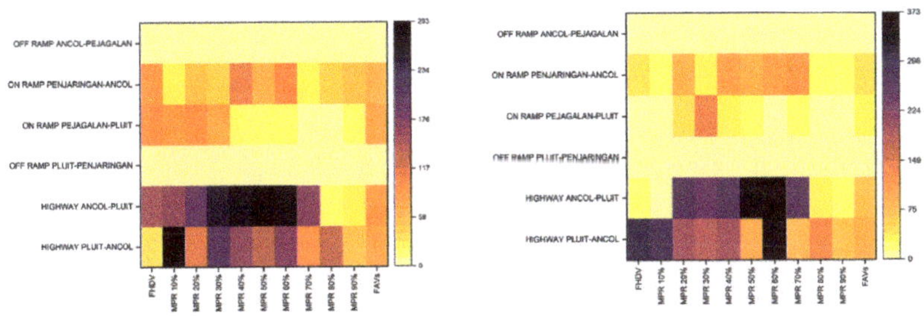

Figure 15. Graph comparing the fuel consumption of aggressive (**left**) and cautious (**right**) autonomous vehicles.

5. Conclusions

This study highlights the transformative potential of autonomous vehicles (AVs) in optimizing traffic flow and reducing environmental impacts. Aggressive AVs consistently outperform cautious AVs across key metrics such as level of service (LOS), average speed, traffic volume, queue delays, emissions, and fuel consumption. Their ability to maintain smaller headways and manage traffic dynamically results in reduced congestion and improved roadway efficiency, particularly at higher market penetration rates (MPRs). Cautious AVs, while improving traffic safety, exhibit limitations in efficiency at lower MPR due to conservative driving behaviors. These vehicles struggle with higher queue delays, longer queue lengths, and increased emissions, especially in mixed traffic conditions. However, their performance improves significantly as the MPR increases, demonstrating their optional behaviour to implementation of autonomous vehicles. This comparative analysis highlights that aggressive driving behaviors surpass cautious ones in achieving optimal traffic flow.

In conclusion, the adoption of aggressive AVs offers significant advantages in improving traffic systems and reducing environmental impacts. However, the transition to fully autonomous systems requires strategic planning, particularly in mixed traffic scenarios with varying MPR. Policymakers, engineers, and urban planners must collaborate to ensure the successful integration of AVs into existing traffic systems, leveraging their strengths to create sustainable, efficient, and safe transportation networks.

Author Contributions: Conceptualization, M.A.; methodology, M.A.; software, M.A.; validation, D.Y.; formal analysis, M.A. and D.Y.; data curation, M.A. and D.Y.; visualization, M.A.; supervision, D.Y. All authors have read and agreed to the published version of the manuscript.

Funding: The research process was conducted by authors funded by the Indonesian Endowment Fund for Education Agency (LPDP) and University of Science & Technology (UST), KICT School.

Institutional Review Board Statement: Not Applicable.

Informed Consent Statement: Not Applicable.

Data Availability Statement: All or some data will be provided upon reasonable request to corresponding author.

Conflicts of Interest: The authors declare no conflicts of interest.

References

1. Wu, C.; Bayen, A.M.; Mehta, A. Stabilizing traffic with autonomous vehicles. In Proceedings of the 2018 IEEE International Conference on Robotics and Automation (ICRA), Brisbane, Australia, 21–25 May 2018; pp. 6012–6018.
2. Azam, M.; Hassan, S.A.; Che Puan, O. Autonomous vehicles in mixed traffic conditions—A bibliometric analysis. *Sustainability* **2022**, *14*, 10743. [CrossRef]
3. Liu, Y.; Guo, J.; Taplin, J.; Wang, Y. Characteristic analysis of mixed traffic flow of regular and autonomous vehicles using cellular automata. *J. Adv. Transp.* **2017**, *1*, 8142074. [CrossRef]
4. Miqdady, T.; de Oña, R.; de Oña, J. Traffic Safety Sensitivity Analysis of Parameters Used for Connected and Autonomous Vehicle Calibration. *Sustainability* **2023**, *15*, 9990. [CrossRef]
5. Hu, X.; Huang, M.; Guo, J. Feature Analysis on Mixed Traffic Flow of Manually Driven and Autonomous Vehicles Based on Cellular Automata. *Math. Probl. Eng.* **2020**, *1*, 7210547. [CrossRef]
6. Ko, W.; Park, S.; So, J.; Yun, I. Analysis of Effects of Autonomous Vehicle Market Share Changes on Expressway Traffic Flow Using IDM. *J. Korea Inst. Intell. Transp. Syst.* **2021**, *20*, 13–27. [CrossRef]
7. Morando, M.M.; Truong, L.T.; Vu, H.L. Investigating safety impacts of autonomous vehicles using traffic micro-simulation. In Proceedings of the Australasian Transport Research Forum 2017 Proceedings, Auckland, New Zealand, 27–29 November 2017; pp. 1–6.
8. Cheng, Y.; Liu, Z.; Gao, L.; Zhao, Y.; Gao, T. Traffic Risk Environment Impact Analysis and Complexity Assessment of Autonomous Vehicles Based on the Potential Field Method. *Int. J. Environ. Res. Public Health* **2022**, *19*, 10337. [CrossRef] [PubMed]
9. Park, S.; Kim, J.; Lee, S.; Hwang, K. Experimental Analysis on control constraints for connected vehicles using Vissim. *Transp. Res. Procedia* **2017**, *21*, 269–280. [CrossRef]
10. Aria, E.; Olstam, J.; Schwietering, C. Investigation of Automated Vehicle Effects on Driver's Behavior and Traffic Performance. *Transp. Res. Procedia* **2016**, *15*, 761–770. [CrossRef]
11. Petrović, Đ.; Mijailović, R.; Pešić, D. Traffic Accidents with Autonomous Vehicles: Type of Collisions, Manoeuvres and Errors of Conventional Vehicles' Drivers. *Transp. Res. Procedia* **2020**, *45*, 161–168. [CrossRef]
12. Abdel-Aty, M.; Ding, S. A matched case-control analysis of autonomous vs human-driven vehicle accidents. *Nat. Commun.* **2024**, *15*, 4931. [CrossRef] [PubMed]
13. Talebpour, A.; Mahmassani, H.S. Influence of connected and autonomous vehicles on traffic flow stability and throughput. *Transp. Res. Part C Emerg. Technol.* **2016**, *71*, 143–163. [CrossRef]
14. Zeidler, V.; Buck, H.S.; Kautzsch, L.; Vortisch, P.; Weyland, C.M. Simulation of autonomous vehicles based on Wiedemann's car following model in PTV vissim. In Proceedings of the 98th Annual Meeting of the Transportation Research Board (TRB), Washington, DC, USA, 13–17 January 2019; pp. 13–17.
15. Azam, M.; Hassan, S.A.; Puan, O.C.; Azhari, S.F.; Faiz, R.U. Performance of autonomous vehicles in mixed traffic under different demand conditions. *Int. J. Automot. Mech. Eng.* **2022**, *19*, 10050–10062. [CrossRef]
16. Shin, S.; Cho, Y.; Lee, S.; Park, J. Assessing Traffic-Flow Safety at Various Levels of Autonomous-Vehicle Market Penetration. *Appl. Sci.* **2024**, *14*, 5453. [CrossRef]
17. Ye, W.; Wang, C.; Chen, F.; Yan, S.; Li, L. Approaching autonomous driving with cautious optimism: Analysis of road traffic injuries involving autonomous vehicles based on field test data. *Inj. Prev.* **2021**, *27*, 42–47. [CrossRef] [PubMed]

18. Park, B.; Qi, H. Microscopic simulation model calibration and validation for freeway work zone network—A case study of VISSIM. In Proceedings of the IEEE Intelligent Transportation Systems Conference, Toronto, ON, Canada, 17–20 September 2006.
19. He, Y. Calibration of Car Following Model with Genetic Algorithm and Particle Swarm Optimization methods. Bachelor's Thesis, Technical University of Munich, München, Germany, 2022.

engineering proceedings

Proceeding Paper

Applying Transformer-Based Dynamic-Sequence Techniques to Transit Data Analysis †

Bumjun Choo and Dong-Kyu Kim *

Department of Civil & Environmental Engineering, Seoul National University, 1, Gwanak-ro, Gwanak-gu, Seoul 08826, Republic of Korea; bjchoo@snu.ac.kr
* Correspondence: dongkyukim@snu.ac.kr

† Presented at the 2025 Suwon ITS Asia Pacific Forum, Suwon, Republic of Korea, 28–30 May 2025.

Abstract

Transit systems play a vital role in urban mobility, yet predicting individual travel behavior within these systems remains a complex challenge. Traditional machine learning approaches struggle with transit trip data because each trip may consist of a variable number of transit legs, leading to missing data and inconsistencies when using fixed-length tabular representations. To address this issue, we propose a transformer-based dynamic-sequence approach that models transit trips as variable-length sequences, allowing for flexible representation while leveraging the power of attention mechanisms. Our methodology constructs trip sequences by encoding each transit leg as a token, incorporating travel time, mode of transport, and a 2D positional encoding based on grid-based spatial coordinates. By dynamically skipping missing legs instead of imputing artificial values, our approach maintains data integrity and prevents bias. The transformer model then processes these sequences using self-attention, effectively capturing relationships across different trip segments and spatial patterns. To evaluate the effectiveness of our approach, we train the model on a dataset of urban transit trips and predict first-mile and last-mile travel times. We assess performance using Mean Absolute Error (MAE) and Root Mean Squared Error (RMSE). Experimental results demonstrate that our dynamic-sequence method yields up to a 30.96% improvement in accuracy compared to non-dynamic methods while preserving the underlying structure of transit trips. This study contributes to intelligent transportation systems by presenting a robust, adaptable framework for modeling real-world transit data. Our findings highlight the advantages of self-attention-based architectures for handling irregular trip structures, offering a novel perspective on a data-driven understanding of individual travel behavior.

Keywords: travel behavior prediction; variable-length sequences; data-driven mobility analysis; first-mile and last-mile travel

1. Introduction

Public transit systems play a critical role in sustaining the mobility of densely populated cities. However, accurately predicting travel behavior surrounding the use of public transit remains a challenge due to the inherent variability in trip structures. As illustrated in Figure 1, each individual trip may consist of a variable number of transit legs, which leads to missing data and inconsistencies when using fixed-length tabular representations. Traditional machine learning approaches struggle with such variability in sequential travel data, often requiring extensive preprocessing and imputation to accommodate incomplete records. For example, methods such as ARIMA [1], Kalman filters [2], and even more recent techniques like random forest [3] generally assume fixed-length inputs and are thus not well-equipped to handle the irregularities inherent in transit data.

Name	Leg 1	Leg 2	Leg 3	Leg 4	Leg 5
User 1	Bus	Bus	Bus	N/A	N/A
User 2	Subway	Bike	N/A	N/A	N/A
User 3	Subway	Bus	Bus	Bus	N/A
User 4	Train	N/A	N/A	N/A	N/A

Tabular Data
Difficult to Apply Machine Learning Algorithms

User 1	Bus	Bus	Bus	
User 2	Subway	Bike		
User 3	Subway	Bus	Bus	Bus
User 4	Train			

Dynamic Sequence Data
Can be Utilized by
Transformer Architecture

Figure 1. Comparison of tabular and dynamic-sequence travel data.

To address these challenges, we propose a transformer-based dynamic-sequence approach that models transit trips as variable-length sequences. By leveraging the power of attention mechanisms, transformer models enable the flexible representation of each trip, preserving both spatial and temporal context [4]. In our framework, each transit leg is encoded as a token that integrates travel time, a mapped mode indicator, and a 2D positional encoding derived from grid-based spatial coordinates. This novel tokenization not only mitigates the issues arising from missing or incomplete data but also enables the model to capture long-range dependencies across different segments of a trip.

In summary, this paper aimed to demonstrate our findings in the following areas:

- Dynamic-Sequence Modeling: We develop a transformer-based model that effectively handles variable-length transit trip data.
- Travel Data Tokenization: We introduce a tokenization method that integrates travel time, mode information, and grid-based positional encoding, capturing complex spatiotemporal patterns inherent in travel data.
- Independent Regression Pipelines: We independently predict first-mile and last-mile travel times, evaluated with Mean Absolute Error (MAE) and Root Mean Squared Error (RMSE), demonstrating superior performance compared to traditional approaches.

The next section describes the dataset characteristics and the preprocessing techniques used to handle variable-length sequences and missing data. Following that, we explain our methodology, including the token generation process, transformer model architecture, and training procedures. Finally, the Results and Implications section presents our findings and discusses their significance for modern travel data analysis.

2. Data

First- and last-mile travel data can be considered microscopic, as they record individual movements. Thus, the collection of such data must adhere to stringent privacy regulations. For this research, *2021 Individual Travel Survey Data* provided by the Korea Transport Database [5] was utilized. This dataset contained encrypted survey data for over 350,000 individual single-purpose travel records. Data such as travel purpose, travel modes, and origin-destination (OD) coordinates for each transit leg were recorded.

Among this data, only travel records that had their OD coordinates in Seoul, South Korea, were considered.

Detailed spatial data such as OD coordinates were aggregated to satisfy privacy requirements and enable standardized positional encoding for model implementation. Using administrative boundary data from Statistics Korea [6], we divided the study area into a 13 × 11 grid of 3 km × 3 km cells. Figure 2 shows this spatial aggregation scheme: each cell is assigned a positional label from (1,1) to (13,11), for a total of 143 cells.

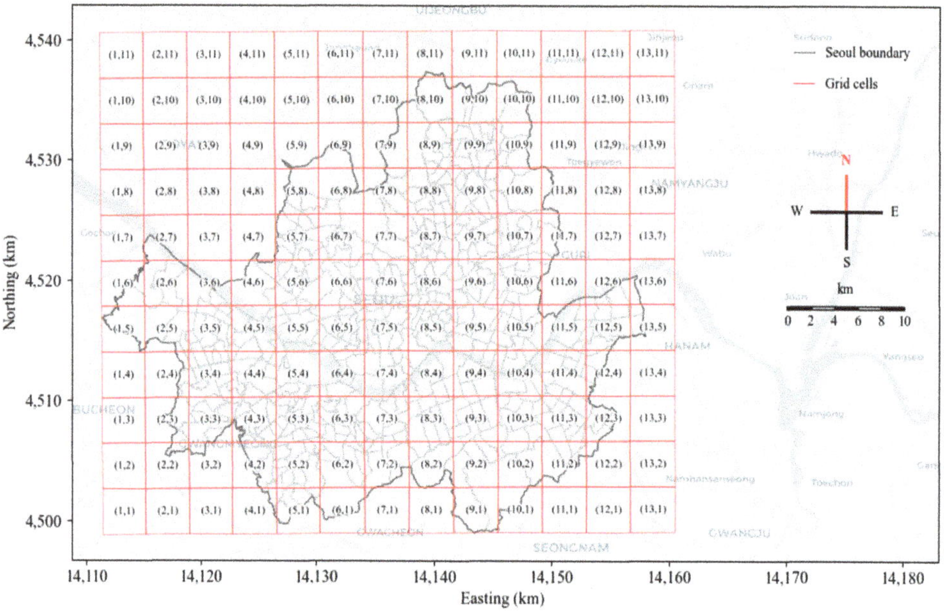

Figure 2. Spatial aggregation of data into 3 km × 3 km cells.

The data was then split into three separate tables:

- First-mile (FM) data: data regarding the first transit leg before public transit use.
- Public Transit (PT) data: data recording one or more consecutive public transit uses.
- Last-mile (LM) data: data relevant to the final transit leg after public transit use.

Specifically, PT data was processed to reveal detailed spatiotemporal characteristics regarding each transit leg, and the target for prediction was set as FM and LM travel time data. The methodology used will be illustrated in the following section.

3. Methodology

Our approach to predicting first-mile and last-mile travel times from public transit data is based on a transformer architecture specifically designed to handle the inherent variability in transit trip records. The methodology is organized into three key components: dynamic-sequence generation through tokenization and positional encoding, a transformer model architecture for regression, and a comprehensive training and evaluation procedure.

In the tokenization process, each trip is converted into a sequence of tokens, with each token representing critical information from one leg of the journey. Each token is constructed by combining three elements: the travel time, the travel mode, and a positional encoding generated using sinusoidal functions. This design, illustrated in Figure 3, ensures that the model captures both the temporal and spatial aspects of each transit leg.

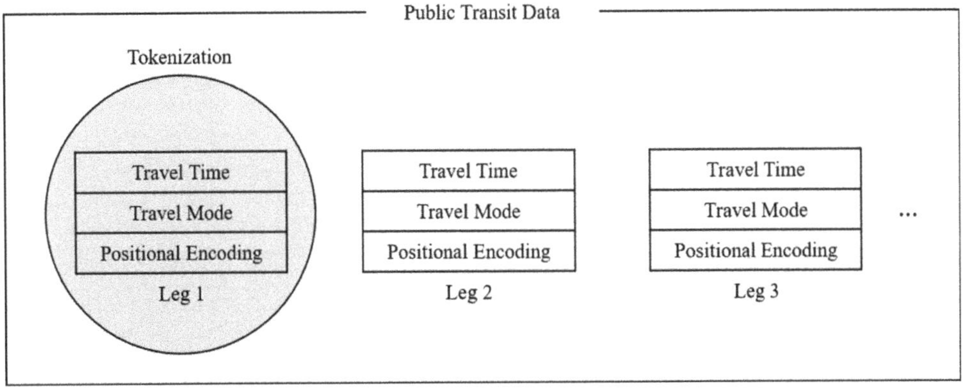

Figure 3. Tokenization of public transit data.

Because transit data often suffers from missing spatiotemporal information [7], our method also incorporates special tokens. Start tokens are generated from the starting grid coordinates of the first public transit leg, while end tokens are generated from the ending grid coordinates of the last leg. When grid data is missing, default values are used to preserve the dynamic-sequence format. This comprehensive tokenization strategy allows the transformer architecture to leverage all available information for data-driven prediction.

Transformer-based regressors leverage the power of self-attention to capture long-range dependencies and complex interactions within data [8]. In our case, the transformer model is built upon a stack of transformer encoder layers, each of which integrates a multi-head self-attention mechanism with a position-wise feed-forward network. Residual connections and layer normalization are used throughout to ensure stable gradients and robust learning.

Moreover, as our input sequences are of variable length, we introduce a masked mean pooling operation following previous literature [9]. This mechanism computes the average representation over only the valid, non-padded tokens, resulting in a fixed-dimensional vector that encapsulates the entire trip.

This pooled vector is then passed through a regression head, which consists of one or more dense layers with non-linear activations and dropout for regularization. The final output layer produces a single continuous value corresponding to the predicted travel time, whether for the first-mile or last-mile segment. The modular nature of this design allows us to train independent models for both first- and last-mile predictions, ensuring each model is finely tuned to its specific target.

4. Results and Implications

ARIMA was employed as a benchmark to evaluate our model because it represents a well-established, classical statistical approach for time series forecasting. Although ARIMA models are traditionally limited to univariate predictions and require extensive preprocessing to handle non-stationary data, they have long served as a baseline in many forecasting studies. By comparing our transformer-based predictions with those of an ARIMA model, we can objectively assess the improvements offered by our advanced, deep learning approach in capturing complex spatiotemporal patterns inherent in transit data.

As can be seen in Table 1, transformer-based models outperformed their ARIMA counterparts for both first-mile and last-mile travel time predictions. Specifically, the transformer model for first-mile travel time (T_FMTT) achieved an MAE of 3.2888 min and an RMSE of 5.1985, while the best performing ARIMA model found through a grid search for the first-mile travel time (A_FMTT) recorded higher errors, with an MAE of 4.1599 min and an RMSE of 7.5323. Similarly, for the last-mile travel time prediction, the transformer model (T_LMTT) yielded an MAE of 4.9373 min and an RMSE of 8.1168, compared to the best performing ARIMA model (A_LMTT), which had an MAE of 7.1521 min and an RMSE of 10.4880. These results show that the transformer model improved MAE by approximately 20.94% for the first-mile travel time prediction and approximately 30.96% for the last-mile travel time prediction, suggesting that the transformer-based approach is more effective at capturing the underlying spatiotemporal dynamics in the transit data, leading to more accurate predictions.

Table 1. First- and last-mile prediction results.

	T_FMTT	T_LMTT	A_FMTT	A_LMTT
MAE	3.2888	4.9373	4.1599	7.1521
RMSE	5.1985	8.1168	7.5323	10.4880

The implications of our findings are significant for public transit systems. Enhanced prediction accuracy enables more precise scheduling, reduces passenger waiting times, and improves resource allocation, all of which contribute to greater operational efficiency. Transit agencies can leverage these insights to develop more resilient and adaptive service plans, ultimately increasing passenger satisfaction and fostering more effective urban mobility management. Furthermore, well-trained models capable of accurately forecasting individual travel characteristics pave the way for integrating these predictions with larger datasets. For example, data fusion techniques [10] could be utilized to generate more granular and detailed travel information at a larger scale than was previously available.

Author Contributions: Conceptualization, B.C.; data curation, B.C.; writing—original draft preparation, B.C.; writing—review and editing, D.-K.K.; supervision, D.-K.K. All authors have read and agreed to the published version of the manuscript.

Funding: This work is financially supported by Korea Ministry of Land, Infrastructure and Transport (MOLIT) as Innovative Talent Education Program for Smart City.

Institutional Review Board Statement: Not applicable.

Informed Consent Statement: Not applicable.

Data Availability Statement: Data used for analysis is provided by Korea Transport Database (KTDB) and available upon request online: www.ktdb.go.kr/www/contents.do?key=202 (access on 3 August 2025).

Conflicts of Interest: The authors declare no conflict of interest.

References

1. Suwardo, W.; Napiah, M.; Kamaruddin, I. ARIMA models for bus travel time prediction. *IEM J.* **2010**, *71*, 49–58.
2. Liu, H.; Zuylen, H.V.; Lint, H.V.; Salomons, M. Predicting urban arterial travel time with state-space neural networks and Kalman filters. *Transp. Res. Rec.* **2006**, *1968*, 99–108.
3. Cheng, L.; Chen, X.; De Vos, J.; Lai, X.; Witlox, F. Applying a random forest method approach to model travel mode choice behavior. *Travel Behav. Soc.* **2019**, *14*, 1–10.
4. Hong, Y.; Martin, H.; Raubal, M. How do you go where? improving next location prediction by learning travel mode information using transformers. In Proceedings of the 30th International Conference on Advances in Geographic Information Systems (SIGSPATIAL '22), Seattle, WA, USA, 1–4 November 2022; pp. 1–10.
5. Korea Transport Database (KTDB). Available online: www.ktdb.go.kr/www/contents.do?key=202 (accessed on 10 July 2024).
6. Statistics Korea (KOSTAT). Korea Administrative District Boundary Data. Available online: https://sgis.kostat.go.kr/view/pss/openDataIntrcn (accessed on 10 July 2024).

7. Park, J.-H.; Kim, S.-G.; Cho, C.-S.; Heo, M.-W. The study on error, missing data and imputation of the smart card data for the transit OD construction. *J. Korean Soc. Transp.* **2008**, *26*, 109–119.
8. Grigsby, J.; Wang, Z.; Nguyen, N.; Qi, Y. Long-range transformers for dynamic spatiotemporal forecasting. *arXiv* **2021**, arXiv:2109.12218.
9. Hou, L.; Geng, Y.; Han, L.; Yang, H.; Zheng, K.; Wang, X. Masked Token Enabled Pre-Training: A Task-Agnostic Approach for Understanding Complex Traffic Flow. *IEEE Trans. Mobile Comput.* **2024**, *23*, 11121–11132.
10. Kusakabe, T.; Yasuo, A. Behavioural data mining of transit smart card data: A data fusion approach. *Transp. Res. Part C Emerg.* **2014**, *46*, 179–191.

MDPI AG
Grosspeteranlage 5
4052 Basel
Switzerland
Tel.: +41 61 683 77 34

Engineering Proceedings Editorial Office
E-mail: engproc@mdpi.com
www.mdpi.com/journal/engproc

www.ingramcontent.com/pod-product-compliance
Lightning Source LLC
Chambersburg PA
CBHW062107220926
51850CB00010B/2921
* 9 7 8 3 7 2 5 8 5 2 3 5 2 *